Proceedings of 1st GENZERO Workshop

Martin Andreoni · Shreekant Thakkar
Editors

Proceedings of 1st GENZERO Workshop

Revolutionizing Autonomous Systems Security with Generative AI, and Large Language Models for Zero Trust Architecture

Editors
Martin Andreoni
Technology Innovation Institute
Masdar, Abu Dhabi, United Arab Emirates

Shreekant Thakkar
Technology Innovation Institute
Masdar City, Abu Dhabi, United Arab Emirates

ISBN 978-981-95-1049-8   ISBN 978-981-95-1050-4 (eBook)
https://doi.org/10.1007/978-981-95-1050-4

© The Editor(s) (if applicable) and The Author(s) 2026. This book is an open access publication.

Open Access This book is licensed under the terms of the Creative Commons Attribution 4.0 International License (http://creativecommons.org/licenses/by/4.0/), which permits use, sharing, adaptation, distribution and reproduction in any medium or format, as long as you give appropriate credit to the original author(s) and the source, provide a link to the Creative Commons license and indicate if changes were made.
The images or other third party material in this book are included in the book's Creative Commons license, unless indicated otherwise in a credit line to the material. If material is not included in the book's Creative Commons license and your intended use is not permitted by statutory regulation or exceeds the permitted use, you will need to obtain permission directly from the copyright holder.
The use of general descriptive names, registered names, trademarks, service marks, etc. in this publication does not imply, even in the absence of a specific statement, that such names are exempt from the relevant protective laws and regulations and therefore free for general use.
The publisher, the authors and the editors are safe to assume that the advice and information in this book are believed to be true and accurate at the date of publication. Neither the publisher nor the authors or the editors give a warranty, expressed or implied, with respect to the material contained herein or for any errors or omissions that may have been made. The publisher remains neutral with regard to jurisdictional claims in published maps and institutional affiliations.

This Springer imprint is published by the registered company Springer Nature Singapore Pte Ltd.
The registered company address is: 152 Beach Road, #21-01/04 Gateway East, Singapore 189721, Singapore

If disposing of this product, please recycle the paper.

Preface

The 2024 GENZERO Workshop, hosted by the Technology Innovation Institute (TII) Secure Systems Research Center (SSRC), convened leading researchers, industry pioneers, and policymakers from November 12 to 13, 2024. Over the course of two days, participants explored how zero-trust security—originally conceived for network defense—can be adapted to the ever-expanding world of AI-driven autonomous systems. A core insight was the urgent need to leverage Generative AI and LLMs to build a robust, end-to-end platform that enforces security, safety, and resilience at all operational layers.

One pivotal focus was the progressive integration of AI into physical robotics, including UAVs (Unmanned Aerial Vehicles), UGVs (Unmanned Ground Vehicles), and, later this year, humanoid robots. When AI directs machinery in the physical realm, it can cause serious harm to humans if not underpinned by strict zero-trust measures. This includes secure communications between command-and-control (C2) systems and swarm units and secure runtime assurance to verify ongoing reliability. Such rigor is paramount when Generative AI detects jamming threats or orchestrates dynamic channel/band switching. As AI's physical footprint grows, so does the imperative to ensure the complete trustworthiness of these robotic platforms.

Goals and Significance

Zero-Trust Evolution for Autonomous Systems

- **Core Objective**: Extend zero-trust from straightforward network paradigms to encompass hardware, software, data pipelines, and AI models within physical robotics.
- **Safety & Resilience**: Embed continuous validation and secure runtime assurance protocols to prevent catastrophic outcomes where robots—UAVs, UGVs, or humanoids—malfunction or fall under adversarial control.

Harnessing GenAI and LLMs

- **Enabling Technologies**: Generative AI and LLMs enable autonomous systems to interpret real-world data, adapt mission parameters, and respond to emerging threats.
- **Balancing Act**: Achieve strong performance without compromising encryption, authentication, and anomaly detection mechanisms—cornerstones of a zero-trust security posture.

Platform-Oriented Vision

- **Platform Approach**: Advocate an ecosystem linking AI-based controllers, secure communications, and robust ML algorithms, all continuously monitored for reliability.
- **Secure Runtime Assurance**: Deliver near real-time oversight of system integrity, ensuring that as tasks scale from ground to air—and eventually to humanoid robotics—, trust remains uncompromised.

Trusted Standard LLM Models & Explainable AI

- **Trust & Transparency**: Many organizations rely on commercial LLMs; the workshop underscored the need to rigorously scrutinize their origins and data.
- **Explainability**: In physical AI deployments, XAI (explainable AI) is crucial. Operators and regulatory bodies must understand why autonomous systems behave as they do, especially in sensitive or high-stakes situations.

Submissions, Review Process, and Selection

- Submissions: 42
- Accepted: 30
- Acceptance Rate: 71%
- Review Method: Double-blind review based on novelty, technical depth, and practical relevance.

This level of scrutiny ensured a compelling mix of foundational research and real-world case studies relevant to zero-trust principles in physical AI.

Workshop Structure and Key Challenges

Six Major Challenges

Workshop sessions explored the multifaceted demands of secure autonomy, from on-device hardware resilience to moral and legal considerations of deploying robots around humans.

Interactive Poster Sessions

Researchers demonstrated early prototypes and proof-of-concept results, engaging in spirited discussions. A peer voting process informed resource allocation for especially innovative ideas.

Invited Talks

Five distinguished experts provided comprehensive insights into the evolving synergy between AI, robotics, and security:

- **Prof. Edward A. Lee** (UC Berkeley) – *"Certainty Or Intelligence"*
- **Prof. Ibrahim Habli** (University of York) – *"Autonomous AI: Who Will Be Responsible for Assuring Its Safety?"*
- **Prof. Antoine Cully** (Imperial College London) – *"Adaptive Machines: Making Adaptive and Resilient Robots with Generative AI and Reinforcement Learning"*
- **Dr. Elahe Arani** (TU Eindhoven & Wayve) – *"E2E Autonomous Driving: The Road to Scalable and Safe Embodiment"*
- **Prof. Morteza (Mory) Gharib** (Caltech) – *"Expanding Human Perception and Real-Time Insights"*

Panel Discussions

Three expert panels addressed crucial aspects of AI security, autonomy, and zero-trust:

- **Panel 1**
    - **Moderator**: Prof. Ernesto Damiani (Khalifa University)
    - **Topic**: *"Can GenAI Truly Overcome the Technical Barriers to Autonomous Systems Recovery?"*
    - **Panelists**: Prof. Peter Pietzuch (Imperial College), Prof. Wenke Lee (Georgia Tech), Prof. Chan Yeob Yeun (Khalifa University), Prof. Edward Lee (UC Berkeley)

- **Panel 2**
    - **Moderator**: Prof. Sandro Pinto (University of Minho)
    - **Topic**: *"Are Zero-Trust Frameworks Merely a Technological Placebo in Autonomous AI Systems Security?"*
    - **Panelists**: Prof. Duen Horng (Polo) Chau (Georgia Tech), Prof. Martin Saska (Czech Technical University), Prof. Muhammad Shafique (NYUAD), Prof. Ibrahim Habli (University of York)

- **Panel 3**
    - **Moderator**: Prof. Martin Saska (Czech Technical University in Prague)
    - **Topic**: *"Will Autonomous Edge Devices Ever Achieve Reliable Decision-Making Without Human Oversight?"*
    - **Panelists**: Prof. Davide Rossi (University of Bologna), Prof. Elahe Arani (TU Eindhoven & Wayve), Prof. Antoine Cully (Imperial College London), Prof. Morteza Gharib (Caltech)

Emerging Themes and Research Frontiers

Holistic Zero-Trust Platforms for Physical Robotics

- **AI-Driven C2**: Advanced command-and-control architectures, themselves AI agents, must incorporate secure runtime assurance to maintain reliability across swarms or fleets of UAVs, UGVs, and humanoids.
- **Communications as an AI Agent**: Generative AI's role in detecting jamming and automatically switching channels/bands exemplified new ways to defend mission-critical data links.

Transparency from AI Vendors

- **Open Datasets & Training Protocols**: Major concerns persist regarding proprietary LLMs with hidden or unclear data origins.
- **Zero-Trust Open-Source Models**: Workshop members called for auditable data pipelines so that integrators can confidently deploy AI on physical robots without unseen risks.

Validating Standard LLMs for Mission-Critical Use

- **Challenges**: Many standard or pre-trained AI models aren't explicitly designed for real-world robotic tasks. Ensuring they meet zero-trust benchmarks is an uphill battle.
- **Opportunities**: Tools for XAI and runtime verification can help unify mainstream LLMs with advanced robotics, fueling safer deployments in public and commercial domains.

Secure Runtime Assurance

- **Key Imperative**: In physically embodied AI, continuous runtime checks are indispensable for detecting anomalies or malicious interference—particularly in high-stakes environments.
- **Next Steps**: Codify industry standards and best practices for monitoring, detection, and rapid response when AI-driven robots deviate from expected behavior.

Future of GenAI in Multi-Agent, Resilient Robotics

- **Adaptive & Self-Repairing**: Generative AI can help robots adapt rapidly to mechanical failures or environmental hazards, provided robust security frameworks guide these adjustments.
- **Scaling to Humanoids**: With humanoid robots on the near horizon, the real-world consequences of AI decisions become more substantial, driving home the necessity of zero-trust from day one.

Looking Ahead: Recommendations and Next Steps

Establishing Transparent Vendor Frameworks

- **Standardized Disclosures**: Urge AI model providers to document training data, methodologies, and alignment strategies for mission-critical robotics.
- **Third-Party Audits**: Independent evaluators can confirm compliance with zero-trust and secure runtime assurance mandates, fostering trust in physical AI solutions.

Benchmarking Zero-Trust & XAI

- **New Metrics**: Develop specialized tests tailored for physical autonomy—including swarms, humanoids, and multi-agent tasks—measuring resilience, transparency, and runtime reliability.
- **Community Repositories**: Encourage data sharing on real-world attempts at AI-driven jamming detection, channel-hopping strategies, and secure humanoid operation.

Regulatory & Ethical Guidelines

- **Liability & Accountability**: As robots enter human spaces, define who is legally liable for AI-driven harm or misuse.
- **Ethical Oversight**: Require safe boundaries around how advanced AI is deployed in public, guaranteeing that technology companies, researchers, and regulators share responsibility.

Interdisciplinary Collaboration

- **Joint R&D Efforts**: Bring together roboticists, cybersecurity experts, AI ethicists, and legal scholars to co-design multi-layered solutions for emergent humanoid robotics.
- **Global Partnerships**: Forge international agreements that harmonize zero-trust standards, ensuring consistent, transparent practices across borders.

Concluding Remarks

The 2024 GENZERO Workshop demonstrated that zero-trust principles must be deeply interwoven into AI-driven robotics—from UAVs and UGVs to future humanoids—to avert potentially serious human harm. By harnessing Generative AI, LLMs, and continuous runtime assurance, participants envisioned an end-to-end secure, resilient, and safe environment where robots can perform critical tasks under constant verification.

Nonetheless, key debates remain around verifying standard LLM models, enforcing vendor transparency, integrating explainable AI, and codifying liability in human-robot scenarios. These priority areas will guide the research landscape into GENZERO 2025 and beyond, ensuring that as physical AI takes on ever-greater autonomy, it also upholds the highest standards of safety, ethics, and zero-trust security.

We extend our heartfelt thanks to the authors, reviewers, and organizing team members for their contributions. Special thanks to all speakers and attendees whose involvement was crucial to the workshop's success.

GENZERO 2024 Workshop Chairs

Martin Andreoni
Shreekant Ticky Thakkar

Contents

Challenge 1: – Resilient Edge AI for Hierarchical Drone Swarms

RESiLIENT: A Neural-Symbolic Resilient Threat-Response Framework for Large-Scale Hierarchical Swarms 3
   Yifan Guo, Kartik A. Pant, Sounghwan Hwang, Vishnu Vijay, James M. Goppert, and Inseok Hwang

Robust and Efficient AI-Based Attack Recovery in Autonomous Drones 12
   Diego Ortiz Barbosa, Luis Burbano, Siwei Yang, Zijun Wang, Alvaro A. Cardenas, Cihang Xie, and Yinzhi Cao

Autonomous Drone Swarms Using Lightweight LLMs 21
   A. Azzouni and G. Pujolle

Quantization-Based Privacy Preservation for Federated Learning in the Sky 32
   Lamees M. Al Qassem, Maurizio Colombo, Ernesto Damiani, Rasool Asal, Al Anoud Almemari, and Yousof Alhammadi

Human-Drone Swarm Collaboration Using LLMs: Case Study on DRL-Based Anti-jamming 38
   Abubakar S. Ali, Shimaa Naser, Omar Alhussein, Sami Muhaidat, and Ernesto Damiani

Challenge 2: – Swarm-Level Threat Intelligence and Response System

Multi-modal Swarm Intelligence for Secure UAV Missions 47
   Yunming Xiao, Mushtari Sadia, and Ang Chen

SHIELD: Swarm-Enabled Hierarchical Intelligent Edge Defense for Drone Swarms 56
   Tamoghna Sarkar and Bhaskar Krishnamachari

Swarming Tight Interactions for Achieving Resistibility of Large Robotic Systems in Real-World Conditions 66
   Jiri Horyna and Martin Saska

RoboMesh: Swarm-Based Orchestration for Secure Multi-robot Systems 72
   Reyhaneh Rabaninejad and Antonis Michalas

Challenge 3: – Adaptive Learning for Evolving Drone Operation

Trusting Data Updates to Drone-Based Model Evolution 81
  Marco Anisetti, Claudio A. Ardagna, Nicola Bena, Ernesto Damiani,
  Chan Yeob Yeun, and Sangyoung Yoon

Adaptive Machines: Making Adaptive and Resilient Robots
with Generative AI and Reinforcement Learning 90
  Antoine Cully and the Adaptive and Intelligent Robotics Lab

Improving Resilience, Security, and Safety of Drones Through
HTM-Based Adaptive Learning 98
  Avi Mendelson, Leonid Azriel, and Adam Ghadban

Robustness of Visual-Based Aerial Navigation to Real-World Adversarial
Attacks 106
  Yaniv Nemcovsky, Chaim Baskin, and Avi Mendelson

Challenge 4: – Enhanced Communication and Active Protection Framework

Enhanced Security and Coordination Framework for UAV Swarms Using
Heterogeneous Communication Networks 117
  Daniel Bonilla Licea, Giuseppe Silano, and Martin Saska

ZETInChat: Zero Trust Infrastructure with Dynamic Service Deployment
via Chatbot in Mesh Networks 124
  Guilherme Nunes Nasseh Barbosa and Diogo Menezes Ferrazani Mattos

Securing AI with AI: Novel Framework for Drone Communication Security ... 131
  Andrea Bastoni, Rodolfo Pellizzoni, Miguel Costa, Emanuele Parisi,
  Francesco Barchi, Andrea Acquaviva, and Sandro Pinto

RL-Enhanced LLMs and Rechargeable Jamming Mines: Achieving
Zero-Trust Security for Hierarchical Drone Swarms 138
  Muhammad Shahzad Arif, Sami Muhaidat, and Paschalis C. Sofotasios

Challenge 5: – Human-Swarm Collaboration Interface for Enhanced Drone and Swarm Resilience

Symbolic Constraint-Solving Capabilities of Transformer Large Language
Models 147
  Leyan Pan, Chris Esposo, Jacob Abernethy, Vijay Ganesh, and Wenke Lee

Cognitive-Aware Multi-modal Human-Swarm Interface for Optimal
Collaboration 155
    Sooyung Byeon, Joonwon Choi, and Inseok Hwang

Towards GenAI-Empowered Unmanned Traffic Management with Zero
Trust 165
    Mohammad Atrouz, Abdulhadi Shoufan, and Fayaz Mohamed Haneefa

Adaptive Resilient Swarming Using Attention and Reinforcement Learning 172
    Robert Penicka and Martin Saska

Challenge 6: – Cloud Security and Privacy

Protecting the Intellectual Property of QNNs at the Deep Edge 181
    Miguel Costa, Tiago Gomes, and Sandro Pinto

A Zero-Trust Hardware Platform for LLMs and Generative AI Edge
Applications 188
    Francesco Restuccia, Davide Rossi, and Ryan Kastner

Burning Fetch Execution: A Framework for Zero-Trust Multi-party
Confidential Computing 196
    Shahin Roozkhosh, Bassel El Mabsout, Cristiano Rodrigues,
    Patrick Carpanedo, Denis Hoornaert, Su Min Tan, Benjamin Lubin,
    Marco Caccamo, Sandro Pinto, and Renato Mancuso

Zero-Trust Secure System and Communication Architecture to Support
LLMs on the Edge Cloud Continuum (LLM-EC2) 206
    Kanad Basu and Ifana Mahbub

Predictive Maintenance System for Enhancing Chip Reliability
and Resiliency in UxVs 216
    Freddy Gabbay

Author Index 227

Challenge 1: – Resilient Edge AI for Hierarchical Drone Swarms

RESiLIENT: A Neural-Symbolic Resilient Threat-Response Framework for Large-Scale Hierarchical Swarms

Yifan Guo[✉], Kartik A. Pant, Sounghwan Hwang, Vishnu Vijay, James M. Goppert, and Inseok Hwang

Purdue University, West Lafayette, IN 47907, USA
{guo781,kpant,hwang214,vvijay,jgoppert,ihwang}@purdue.edu
https://sites.google.com/view/fdchsl/home

Abstract. Achieving resilience has become an emerging challenge for large-scale swarm autonomy, entailing both the identification of unforeseen events and the recovery of affected systems from such events. The complexity of modern autonomous systems introduces new system vulnerabilities to adversarial attacks on both physical- and cyber-systems. These vulnerabilities, coupled with the intricate interconnection of large-scale swarm systems, make swarm-level resiliency difficult to obtain. A simple failure in one local system can cascade to others, leading to catastrophe. In this poster, we highlight how neural-symbolic concepts, combining the best of machine learning and control theory in a unified framework, can enhance the resilience of hierarchical swarm operations. Through several illustrative examples, we showcase the advantages of neural-symbolic approaches in two broad categories: *proactive* and *reactive* strategies for resilient hierarchical swarms. With strong interpretability, our approaches collectively achieve resilient planning, continuous evolution, and swift re-organization against unknown threats and anomalies.

Keywords: resilient robotic swarms · neural-symbolic AI · hierarchical robotic swarms

1 Research Motivation

Robotic Swarms (RSs) have been widely utilized for their capability to tackle complex and repetitive tasks, including package delivery [1], surveillance [2], etc. Recently, these systems have been utilized for more strategic and tactical purposes, e.g., the use of Unmanned Aerial/Ground Vehicles (UAVs/UGVs) to enhance situational awareness in emergencies and aid law enforcement agencies. However, most existing swarm deployments are homogeneous and rely on commands from human administrators or operation centers [3]. This homogeneous and centralized nature constrains the size and capability of RSs as manual decision-making is slow and inefficient with limited computational resources.

These existing swarm architectures are also incapable of continuous evolution and adaptation to dynamic environments, which limits their usage to relatively confined spaces and controlled scenarios. To improve efficiency and reliability, researchers traditionally rely on non-learning methods, which are based on, among others, control, graph, and game theories [4,5]. These methods provide solid theoretical foundations and transparent interpretability. However, they usually require precise modeling of the system [6] and tend to simplify the dynamics of complex systems [7], which limits their adaptability to evolving environments. The computation burden of many non-learning algorithms explodes as the size of the problem increases, especially those regarding combinatorial optimization problems [8]. Most importantly, these methods also do not provide any systematic evolution mechanism that can enable the swarm to learn from its experience. Machine learning-based methods, on the other hand, provide excellent scalability and adaptability while dealing with high-dimensional input data. They outperform non-learning methods in various applications, such as aircraft aerobatic control [9]. However, they are notorious as black boxes and are less trusted for safety-critical missions. Moreover, imposing constraints (e.g., battery capacity, collision avoidance, etc.) or prior knowledge to pure Deep Learning (DL) models is challenging and introduce additional optimization bottlenecks[10].

Thus, we propose a neural-symbolic resilient threat-response framework for large-scale hierarchical swarms, combining the best of machine learning and non-learning theories to comprehend patterns in high-dimensional data while imparting reasoning and interpretability. Figure 3 presents an example neural-symbolic architecture for persistent coverage problem, where a multi-agent system is designated to monitor a region, keeping every grid frequently inspected. The perception neural network efficiently observes the region and generates waypoints. Given waypoints, the MPC module generates trajectories following various constraints (e.g., collision avoidance). Figure 4 is a specific example of resilient persistent coverage. The region of interests is discretized and the shade of blue indicates the "awareness" of a grid. Red circles are obstacles, yellow arrows are agents, and green crosses denote MPC predictions. The MPC module of the circled agent is adjusting its trajectory to avoid the obstacle (Fig. 1 and Fig. 2).

The major features of the proposed framework include:

1. **Hierarchical structure**: Our proposed framework is well suited for hierarchical heterogeneous robotic swarms, which follow the existing cloud-fog-edge structure and consist of various types of robot (e.g., UAVs, UGVs, etc.) [11]. Using our proposed framework, regional data can be aggregated in the fog layer, which relieves the computation and communication burden and enhances system resiliency by reducing the reliance on the cloud. The diversity of robot types and performance enriches the collected data, thus, accelerates the evolution of the swarm.
2. **Neural-symbolic approaches**:
    Our proposed framework employs neural-symbolic methods across most phases of threat response, from network configuration and task assignment to motion planning and control. Powered by the high-level DL model, our neural-

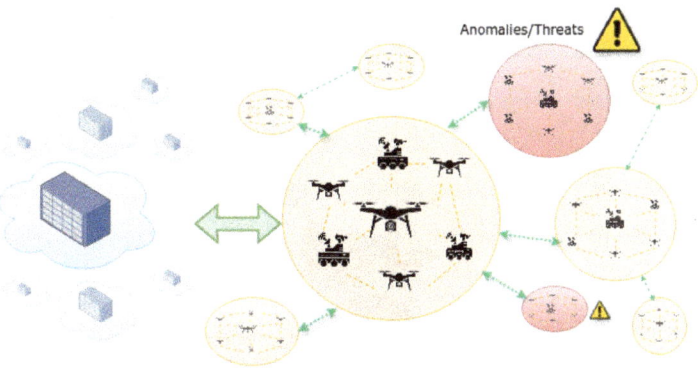

Fig. 1. Large-scale cloud-fog-edge drone swarm architecture under unknown threats or anomalies.

symbolic methods possess superior scalability and adaptability. At the same time, with constraints imposed on low-level non-learning modules, the final outputs are trustworthy and explainable, making the overall system resilient to unknown threats.
3. **Real-time threat detection and continuous learning**: With both proactive and reactive strategies for threat mitigation, our framework greatly enhances the swarm's resilience to unexpected threats and anomalies. Our continuous learning strategies enable collective and task-specific evolution of the entire swarm network, exploiting the hierarchical structure.

2 Proactive Strategies

In this section, we demonstrate the proactive aspect of our framework by leveraging AI-enabled network vulnerability analysis and resilient motion planning with a neural-symbolic scheme to enhance the resilience of the hierarchical swarm.

2.1 Analysis on Network Vulnerability Under Potential Threats

Evaluating network vulnerabilities among edge agents is crucial to enhancing the resilience of large-scale swarm operations. Given the collaborative behaviors between agents (e.g., autonomous vehicles and UAVs) for the missions, improving *network connectivity* is a critical concern that must be addressed. For instance, with respect to surveillance missions, the authors in [12] have indicated that enhanced network connectivity between agents is crucial for real-time coordination, data sharing, and dynamic task allocation, which enables more efficient operations. In other words, it can improve decision-making and situational awareness by allowing the agents to collaborate, share sensor data, and adapt to changing mission conditions. However, in adversarial environments, a swarm's network connectivity may fall below a desirable level due to malicious impacts

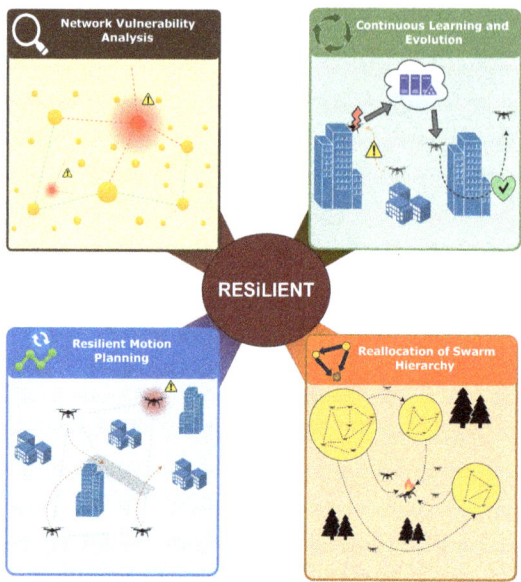

Fig. 2. Our proposed neural-symbolic framework, RESiLIENT, with proactive and reactive strategies to achieve resilient network design and motion planning, and continuous learning and evolution with swift reorganization for real-time distributed threat response in large-scale distributed swarms.

(e.g., cyberattacks and unexpected threats). This can lead to poor mission performance and numerous safety violations, such as crashes and collisions, between agents and/or obstacles.

Additionally, as the number of agents increases, understanding the dynamics of the network through the non-learning model–based approaches (e.g., using Laplacian matrix and geometric properties) may encounter technical challenges due to the curse of dimensionality; and unknown dynamics and effects [13]. Therefore, it is crucial to develop a resilient strategy that allows the swarm to maintain robust network connectivity despite unintentional malicious threats, including Denial-of-Service (DoS) cyberattacks.

Scenario 1: AI–Driven Resilient Coordination of Swarms: In the real world, attackers (those who launch cyberattacks) have limited resources and/or budgets to launch their attack strategies/policies. They will likely target the most critical network links or agents that play a significant role in maintaining important network features, such as network connectivity. With advancements in computational technologies, AI-driven frameworks with neural-symbolic schemes (e.g., pLogicNet and Express GNN) [14] could be promising solutions to address this problem. Considering the coordination and control policies of agents, these AI-driven techniques enable us to predict/estimate which agents (even a group of

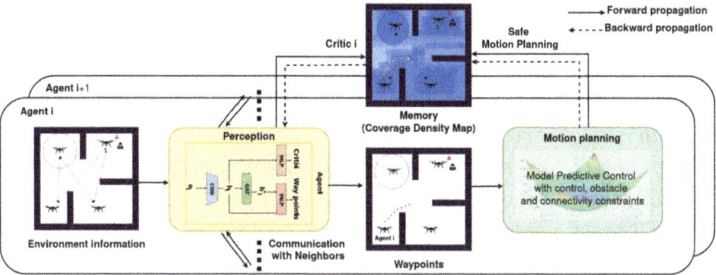

Fig. 3. An example imperative learning model for the persistent coverage problem. Each agent observes its local situation with a perception network, shares the observation with its neighbors, and decides on tentative waypoints for the next time step. With the waypoints, the MPC module generates trajectories considering various constraints (e.g., collision avoidance).

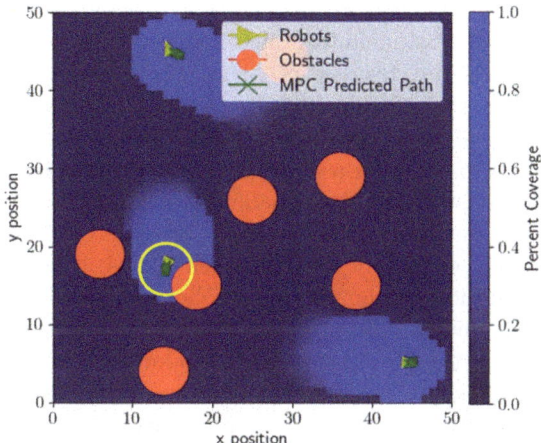

Fig. 4. An illustration of multi-agent persistent coverage through numerical simulations. The background blue indicates the level of awareness, the lighter the better. The local planner based on MPC ensures the agent's safety (circled) even if the high-level learning module provides potentially risky waypoints.

agents) are more vulnerable to cyberattacks (e.g., DoS attacks). Additionally, these methods can classify the level of attack vulnerabilities for each agent, enabling ground stations—entities that manage swarm operations—to create resilient coordination/motion strategies to counter potential (unforeseen) cyber threats. Therefore, our proposed method ensures the resiliency of the hierarchical swarm system.

2.2 Neural-Symbolic Resilient Motion Planning For Swarms

Motion planning is a fundamental operation for almost every advanced mission for robotic swarms. Existing swarm systems employ control-theoretic approaches, such as distributed Model Predictive Control (MPC) and consensus formation control. These approaches have shown great success in generating efficient motions while maintaining state and input constraints (e.g., collision avoidance and battery constraint). However, these methods heavily rely on an exact model of robots in the swarm and environment, which is challenging to achieve in real-world scenarios [15]. Deep learning approaches, on the other hand, are good at environmental perception and dealing with the imperfect model. But they demand a huge amount of expert supervision and hard to interpret.

To this end, we introduce Imperative Learning (IL) [16], where deep learning models are employed to perceive the complex environment for task allocation, and control-theoretic methods are used to handle motion planning.

Scenario 2: Persistent Coverage with Continuous Connectivity: A fundamental problem in the task of RSs, including surveillance and sensor deployment, can be modeled as a persistent coverage problem, where a group of robots should monitor certain areas to maintain a desired coverage level. Traditionally, this problem has been formulated as a constrained optimization problem when a group of robots tries to optimize a *density* function representing the quality of coverage under fuel constraints. These challenges are exacerbated when nefarious actors, e.g., hackers, manipulate the behavior of some of the robots by spoofing sensors or hacking communication channels. Thus, it becomes essential to determine resilient motion planning for each robot in an RS to persistently cover a given environment under collision and connectivity constraints. Connectivity constraints ensure that the RS effectively utilizes locally detected information as an additional redundancy to mitigate the effects of faults and/or cyberattacks. To address this, we propose a neural-symbolic method including three major components: 1) a distributed DNN that perceives its local environment and allocates target waypoints to robots, 2) an MPC–based safe motion planning for each robot given the desired waypoints while satisfying obstacle and connectivity constraints, and 3) a rigid body theory-based anomaly detector, running under the assumption of connectivity.

3 Reactive Strategies

This section focuses on reactive strategies which enable an RS to evolve and adapt continuously from collective experiences. Using an federated learning framework, we demonstrate how to simultaneously achieve global evolution and regional specification. We further describe a reallocation framework that allows a swift reorganization of the swarm's hierarchy once an anomaly or fault is detected.

3.1 Continuous Evolution and Adaptation

For all creatures, it is vital to constantly learn from experience and adapt to environment. This is the same for an RS to stay resilient and efficient in complex environments. RSs also have a huge advantage over natural creatures in that individuals in the swarm can rapidly exchange their experiences and evolve. However, such swift evolution cannot be implemented on RSs with a centralized swarm structure. This is because the centralized structure leads to a dilemma between data processing ability and the manufacturing cost of terminal agents. Moreover, agents can never perform optimally if they are all identical and do not adapt to specific situations. A promising solution to this dilemma is to combine federated learning with a hierarchical swarm structure. With the fog-level agents performing local pre-processing, the cloud server only needs to deal with highly abstract features, which relieves computation and communication workload and ensures security and privacy. This cloud-fog-edge structure is decentralized due to lower-level computations being done at the fog- and edge-level, and does not require constant communication with the central cloud [11]. Encrypted parameters of a global model are downloaded by the agents from a cloud server, and model training is done locally so that sensitive information stays in the local system. Significant research has been done on encryption techniques for federated learning that could be used to provide sufficient privacy without impacting performance, such as homomorphic encryption and secret sharing technology. The encrypted gradient information from the agents is transmitted back to the cloud server for aggregation and update to the global model [17].

Scenario 3: Federated Learning for Hierarchical Swarm The graph structure of the swarm may change frequently due to either mission requirements or unexpected anomalies. Incorporating federated learning with a hierarchical swarm structure, we can leverage the intermediate fog layer between the cloud and edge layers to tailor the global model for each fog. To do so, the fog agent can add a bias to the global model, specifying the task and environment of the local system. When an updated model is sent from the edge agents to the cloud for aggregation, the fog agent can also be used to remove the fog-specific bias.

This method allows edge agents to be quickly reassigned between fogs with negligible performance degradation due to the fog-level agent handling adaptation of the global model to the local swarm's goal and environment.

Anomalies can also be invaluable learning opportunities for the swarm, making the swarm resilient to similar threats in the future. Directly transmitting raw data history is inefficient and insecure, not to mention the damaged agent could shut down very soon after the anomaly happens. Inspired by the idea of one-shot federated learning [18], we propose a method where each edge agent (or its fog instead) periodically distills and stores an *encrypted gradient message*. With local distillation in advance, the size of transmitted data is compressed significantly, yet containing most of the important information. Also, this communication form protects the privacy of raw data since it submits encrypted gradient information to the swarm.

3.2 Reallocation of Swarm Hierarchy

Another critical challenge in reactive threat response is reassigning the swarm hierarchy once agents are identified as faulty or attacked. It leads to a very complex combinatorial optimization problem to select proper agents to form a "special force" and hand over the unfinished work, considering the energy constraints, interventions to current plans, and the heterogeneity of the swarm. Fortunately, active research [19] is going on to obtain near-optimal solutions for large-scale combinatorial optimization problems. Combining the idea of imperative learning, we can simultaneously achieve fast computing, scalability, and interpretability.

Scenario 4: Reorganizing Local Delivery Network: The delivery problem can be formulated as a Multiple Traveling Salesman Problem (MTSP), which is well-known for its NP-hardness. State-of-the-art non-learning approaches can only handle about MTSP with about 150 cities. Based on a novel imperative learning approach [20], we can first decompose the MTSP into a multiple single-agent Traveling Salesman Problem (TSP). Then, we employ a non-learning solver to efficiently solve these sub-problems. This innovative framework can solve the MTSP with dozens of agents and thousands of cities within a few seconds, which can satisfy the demands of any current RS.

References

1. Chen, Z., Alonso-Mora, J., Bai, X., Harabor, D.D., Stuckey, P.J.: Integrated task assignment and path planning for capacitated multi-agent pickup and delivery. IEEE Robot. Autom. Lett. **6**(3), 5816–5823 (2021)
2. Walle, D., Fidan, B., Sutton, A., Yu, C., Anderson, B.D.: Non-hierarchical UAV formation control for surveillance tasks. In: American Control Conference. IEEE, pp. 777–782 (2008)
3. Brambilla, M., Ferrante, E., Birattari, M., Dorigo, M.: Swarm robotics: a review from the swarm engineering perspective. Swarm Intell. **7**, 1–41 (2013)
4. Franco, C., Stipanović, D.M., López-Nicolás, G., Sagüés, C., Llorente, S.: Persistent coverage control for a team of agents with collision avoidance. Eur. J. Control. **22**, 30–45 (2015)
5. Peters, J.R., Wang, S.J., Surana, A., Bullo, F.: Asynchronous and dynamic coverage control scheme for persistent surveillance missions arXiv preprint arXiv:1609.05264 (2016)
6. Wang, N., Lv, S., Liu, Z.: Global finite-time heading control of surface vehicles. Neurocomputing **175**, 662–666 (2016)
7. Selig, M.S.: Real-time flight simulation of highly maneuverable unmanned aerial vehicles. J. Aircr. **51**(6), 1705–1725 (2014)
8. Wang, H., Chen, B., Lin, C., Sun, Y., Wang, F.: Adaptive finite-time control for a class of uncertain high-order non-linear systems based on fuzzy approximation. IET Control Theory Appl. **11**(5), 677–684 (2017)
9. Clarke, S.G., Hwang, I.: Deep reinforcement learning control for aerobatic maneuvering of agile fixed-wing aircraft. In: AIAA Scitech 2020 Forum, p. 0136 (2020)

10. Krishnapriyan, A., Gholami, A., Zhe, S. Kirby, R., Mahoney, M.W.: Characterizing possible failure modes in physics-informed neural networks. Advances in Neural Inf. Process. Syst. **34**, 26 548–26 560 (2021)
11. De Donno, M., Tange, K., Dragoni, N.: Foundations and evolution of modern computing paradigms: Cloud, IoT, Edge, and Fog. IEEE Access, **7**, 150 936–150 948 (2019)
12. Yun, W.J.: Cooperative multiagent deep reinforcement learning for reliable surveillance via autonomous Multi-UAV control. IEEE Trans. Industr. Inf. **18**(10), 7086–7096 (2022)
13. Thapliyal, O., Hwang, I.: Data-driven cyberattack synthesis against network control systems. IFAC-PapersOnLine **56**(2), 8357–8362 (2023)
14. Yu, D., Yang, B., Liu, D., Wang, H., Pan, S.: A survey on neural-symbolic learning systems. Neural Netw. (2023)
15. Moerland, T.M., et al.: Model-based reinforcement learning: a survey. Found. Trends® Mach. Learn. **16**(1), 1–118 (2023)
16. Wang, C., et al.: Imperative learning: a self-supervised neural-symbolic learning framework for robot autonomy. arXiv preprint arXiv:2406.16087 (2024)
17. Li, L., Fan, Y., Tse, M., Lin, K.-Y.: A review of applications in federated learning. Comput. Indust. Eng. **149**, 106854 (2020)
18. Zhou, Y., Pu, G., Ma, X., Li, X., Wu, D.: Distilled one-shot federated learning arXiv preprint arXiv:2009.07999 (2020)
19. Bengio, Y., Lodi, A., Prouvost, A.: Machine learning for combinatorial optimization: a methodological tour d'horizon. Eur. J. Oper. Res. **290**(2), 405–421 (2021)
20. Guo, Y., Ren, Z., Wang, C.: iMTSP: Solving min-max multiple traveling salesman problem with imperative learning arXiv preprint arXiv:2405.00285 (2024)

Open Access This chapter is licensed under the terms of the Creative Commons Attribution 4.0 International License (http://creativecommons.org/licenses/by/4.0/), which permits use, sharing, adaptation, distribution and reproduction in any medium or format, as long as you give appropriate credit to the original author(s) and the source, provide a link to the Creative Commons license and indicate if changes were made.

The images or other third party material in this chapter are included in the chapter's Creative Commons license, unless indicated otherwise in a credit line to the material. If material is not included in the chapter's Creative Commons license and your intended use is not permitted by statutory regulation or exceeds the permitted use, you will need to obtain permission directly from the copyright holder.

Robust and Efficient AI-Based Attack Recovery in Autonomous Drones

Diego Ortiz Barbosa[1]([✉]), Luis Burbano[1], Siwei Yang[1], Zijun Wang[1], Alvaro A. Cardenas[1], Cihang Xie[1], and Yinzhi Cao[2]

[1] University of California Santa Cruz, Santa Cruz, CA 95064, USA
dortizba@ucsc.edu
[2] Johns Hopkins University, Baltimore, MD 21218, USA

Abstract. We introduce an autonomous attack recovery architecture to add common sense reasoning to plan a recovery action after an attack is detected. We outline use-cases of our architecture using drones, and then discuss how to implement this architecture efficiently and securely in edge devices.

Keywords: drone recovery · simplex architecture · Multimodal Large Language Models · Edge Devices

1 Introduction

Autonomous drones or self-driving vehicles are vulnerable to various attacks, such as physical interference affecting sensor readings [19], actuation signals [6], GPS spoofing [15], etc. Such security lapses can cause dangerous consequences in the physical world, such as vehicle crashes [1] or navigation errors [14] that may steer our autonomous vehicle into enemy territory or away from its mission.

To protect these systems, researchers have developed several tools for preventing, detecting, and recovering from attacks. Automatic recovery, the last of these steps, plays a significant role for drones and other autonomous vehicles because if they are attacked, they need to recover quickly to prevent accidents such as crashing or harming humans.

Real-time attack recovery solutions are mainly based on the **simplex architecture**, which consists of two *different* controllers [5,8,21]: One is a nominal controller optimized for performance but without safety guarantees. If an attack is detected, we switch from the nominal controller to the *recovery controller*, a controller that changes the objective of the mission to perform a safety maneuver. These recovery controllers can try to steer the drone to a safe area, even when signals are partially compromised.

1.1 Example

We now illustrate how our work leverages this recovery controller to keep drones safe. Drones must perform different tasks, such as surveillance in adversarial

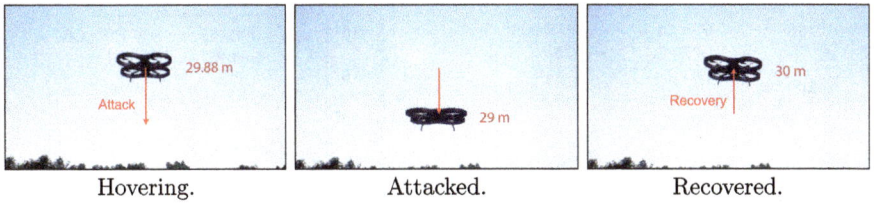

| Hovering. | Attacked. | Recovered. |

Fig. 1. A drone receives false GNSS information, forcing it to lower it's altitude. OPR detects this attacks and returns the drone to a safe altitude.

environments, where attackers might want to land the drone without authorization or produce a crash.

This example is motivated by the RQ-170 UAV incident. In particular, the government of Iran claims they used a cyber-attack to force a U.S. surveillance drone to land in Iranian space [10,16]. In this use case, an attack spoofs GPS signals to make the drone believe it is at a higher altitude than it really is (Fig. 1a). Without any defense, the drone will start descending and eventually land (Fig. 1b). Our attack-recovery mechanism detects the attack (by looking at the inconsistency between control actions and GPS values) and then recovers its original (safe) position by creating virtual sensors: altitude predictions based on physical models (Fig. 1c).

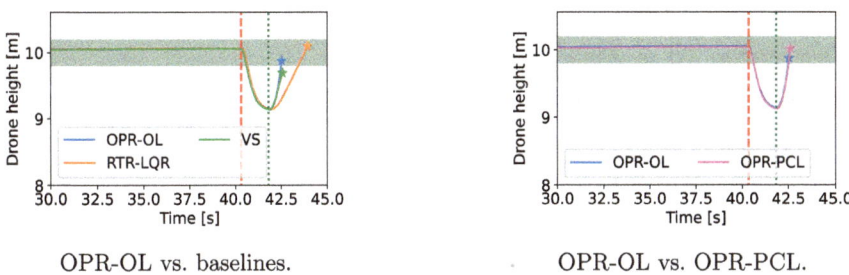

| OPR-OL vs. baselines. | OPR-OL vs. OPR-PCL. |

Fig. 2. Our algorithm (OPR-OL) returns a drone to a safe height (green area) faster and more accurately than previous work. In addition, if we can filter out the malicious sensor and take the input from the remaining sensors, we obtain a Partially Closed Loop (OPR-PCL) algorithm that outperforms slightly our open loop model.

We call our algorithm Optimal Probabilistic Recovery (OPR) [20] and we consider it as Open Loop (OL) if we assume that all sensors are compromised, or Partially Closed Loop (PCL) if we can detect the only signal attacked, and then consider the other sensors as trustworthy. Figure 2a shows that OPR-OL recovers the drone faster than other baselines (Real-Time Recovery with the Linear Quadratic Regulator–RTR-LQR [21] and Virtual Sensors–VS [4]); and Fig. 2b shows how information from non-compromised sensors (OPR-PCL) can improve recovery by landing in the middle of the target set.

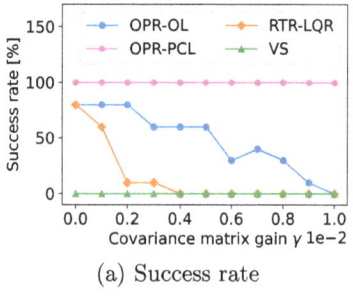 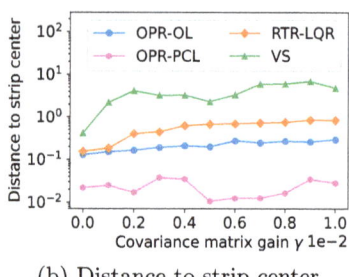

(a) Success rate (b) Distance to strip center

Fig. 3. Success rate and average distance to the target set center with increasing noise for the drone.

OPR-OL and OPR-PCL also outperform the success rates of the baselines (how many attacks are recovered to the target–green–set, in Fig. 3a) and by the distance to the center of the desired target (Fig. 3b).

While these previous efforts can help prevent immediate safety risks, they still require mission planners to identify several parameters before a mission, such as safe destinations to go to (targets) after an attack is detected; and thus they are not adaptable to uncertain conditions and new attacks. In our ongoing work, we plan to address these limitations by leveraging advances in AI.

To make our AI-based attack recovery strategy useful and practical, we argue that we need to solve the following research challenges:

– Design of an AI recovery algorithm.
– Design of efficient and practical algorithms that can run on edge devices or on embedded systems by orchestration with an AI agent in the cloud.
– Design attack-resilient AI agents that are not vulnerable to test-time adversarial example attacks.

2 Challenge 1: Autonomous Recovery

The state-of-the-art automatic attack-recovery mechanisms described in the previous section do not work with dynamic and uncertain environments. For example, these previous methods need precomputed target safe areas where the recovery controller can take the system; however, if these sets are not preloaded in advance, or if the safe zones are not safe during sporadic periods of time, the automatic recovery mechanism will fail.

As the cornerstone of a new era in AI, generative AI (GenAI) models such as Falcon2 [12] and GPT-4 [3] promise to catalyze a profound transformation across numerous sectors of society, providing common sense reasoning in real time to adapt to uncertain and dynamic scenarios. To address the limitations of previous attack recovery systems, we propose a GenAI-Based attack recovery mechanism. Our main insight is to have a hierarchical recovery strategy—At the lower level we will use mathematical control-theory models based on the simplex architecture (as described in the previous section); At a higher level, we will design a generative AI recovery algorithm to provide a common-sense and adaptive recovery plan. Our concept is illustrated in Fig. 4.

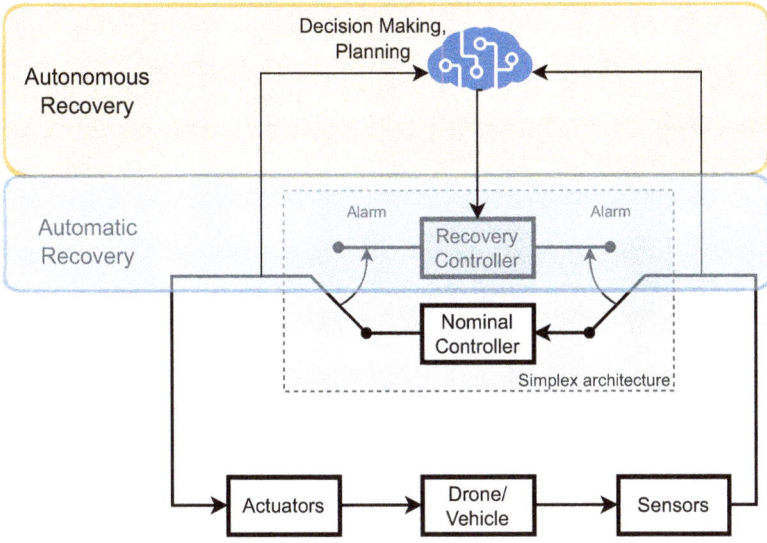

Fig. 4. AI-Based Recovery.

To design an AI-based attack recovery, we need to solve several problems: (1) AI agents need to understand the state of the drone (or vehicle), identify risks, and create action plans. This requires encoding of the state of the physical world into a format that can be understood by the GenAI agent. (2) Identify safety zones dynamically as the mission progresses to give to the lower-level automatic controller, (3) Have a long-term plan for recovering after reaching the target set (e.g., identify if the attack has stopped, when can we engage the nominal controller again, and when do we ask for help from a human operator or other agents).

In particular, we plan to extend our recent work [20] with common sense reasoning to find safe target sets and maneuver toward them after we detect an attack. We define the target sets with two elements: 1) the closed form $T \in \mathcal{T}$, with $\mathcal{T}$ the set of possible forms, and 2) the parameters $\theta \in \Theta$, where Θ the set of all possible parameters. Note that θ depends on the form of the target set T. Then, we denote the set of valid parameters Θ for a target set form T as $\Theta(T)$. For instance, for a drone with n states, the target set can be a strip [20], where we can define that the drone state $x \in \mathbb{R}^n$ is between a range at the end of the recovery. A strip is the intersection between two hyperplanes

$T(\theta) = \{x \in \mathbb{R}^n \mid \theta_1^T x \geq \theta_2 \wedge \theta_1^T x \leq \theta_3\}$, where $\theta_1 \in \mathbb{R}^n$, $\theta_2 \in \mathbb{R}$ and $\theta_3 \in \mathbb{R}$ are the target set parameters. Therefore, we can select θ_1 to define the flying height of the target drone between θ_2 and θ_3.

LLMs can produce the parameters θ. For this, the LLM takes sensor information from the observation set $o \in \mathcal{O}$, the form of the set $T \in \mathcal{T}$, and contextual information such as environmental conditions $c \in \mathcal{C}$ to produce the target set parameters $\theta \in \Theta$. That is, the LLM becomes a function $F : \mathcal{O} \times \mathcal{T} \times \mathcal{C} \to \Theta$.

Using the LLM to define the target set parameters comes with several challenges. First, the LLM may output a target set that is not actually safe. Similarly, the target set may be infeasible; arriving at the target set the LLM generates may be impossible. Therefore, we will work on a verifier mechanism that certifies the safety and feasibility of the target set.

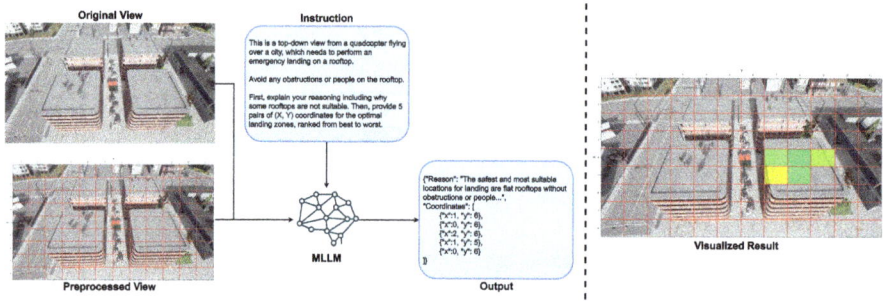

Fig. 5. Multi-modal LLM (MLLM) evaluates the risks and ranks possible safe landing locations.

Figure 5 illustrates a use case of this methodology. After detecting an attack, we ask the LLM to identify a safe area where the drone can land (given the camera feed of the drone). The LLM must decide which of the four buildings the drone should land in. Two of those buildings are crowded with people, while the other two are empty. The LLM needs to identify that empty buildings are safer to land in than crowded ones. The drone's camera feed is first preprocessed with a Cartesian coordinate system added to make it easier to interpret the multi-modal LLM's output. The LLM then predicts several candidate safe landing zones based on both the original and preprocessed views. Grounded decoding is applied in the final stage of the LLM to ensure the output strictly follows the required format. Each predicted landing zone includes coordinates and a "Reason" section to improve prediction accuracy and interoperability.

To improve the reasoning process (and improve the prediction accuracy) of LLMs, we plan to test prompting techniques such as Chain-of-Thought, Self-Consistency, and Self-Reflection. Also, as LLMs may sometimes fail to recognize objects such as buildings and people in the images, a dedicated object detection/semantic segmentation model will be used to recognize objects and then color-code the objects in images as part of a preprocessing process, so that these objects can be easier for LLMs to recognize.

3 Challenge 2: Efficient Edge and Device AI

A critical challenge we face is the reduction of operational latency in GenAI applications. The success of drones in critical missions, such as immediate disaster response or high-speed surveillance operations, is highly dependent on their ability to process and respond to incoming information with minimal delay.

In responding to latency concerns, our aim is to tackle them with recent algorithmic efficiency proposals.

- **Model Distillation:** This technique involves distilling a large language model into a more compact version while retaining the essential features necessary for robust performance. Following recent work [18], our aim is to control the size of multimodal LLM under 0.2 billion parameters, ensuring rapidness without substantial loss in effectiveness.
- **Efficient Mobile Model Design:** Given that traditional transformer architectures exhibit quadratic computational complexity with respect to token length, exploring alternatives such as the Mamba / RWKV model [9,13], which offers linear complexity, is considered advantageous. This modification could significantly reduce computational demand, enabling quicker data processing [2,11].
- **Post-Training Quantization:** Transitioning from floating point precision (fp32 or fp16) to a highly quantization format such as int8 or even a binary version can substantially accelerate model operation [7].

These three strategies can also be used together to further reduce model latency on edge devices, equipping drones with the capability to respond in realtime to diverse and dynamic environmental stimuli. Moreover, to build more capable multimodal LLMs, which requires navigating complex and varied real-world scenarios, we are exploring the following innovative approaches:

- **Learning Every Signal:** To maximize the capabilities of multimodal LLMs, we plan to pioneer diverse tokenization methods aimed at integrating and processing a variety of signals. This strategic development is designed to build a coherent and multifaceted input landscape, encompassing different data types such as images, videos, textual and voice inputs from users, and radar signals. Our objective is to cultivate a robust input framework that significantly boosts the model's capacity to learn and adapt across the spectrum of data encountered in UAV operations.
- **Reinforcement Learning with Human Feedback (RLHF):** We plan to incorporate human feedback into the training loop of our models. This can be achieved by engaging a human copilot who monitors and, if necessary, corrects the UAV's actions during operation. The corrective inputs provided by the human operator are used to reinforce and refine the model's understanding and responses to real-world scenarios. By continuously evaluating and adjusting AI decisions with insights from experienced human experts, our goal is to significantly improve the decision-making capabilities of our systems, especially in complex environments where nuanced judgment and situational awareness are crucial.

4 Challenge 3: Robust GenAI-Based Attack Recovery

We also need strategies to enhance the robustness of GenAI systems to ensure that our recovery system is not abused by attackers.

The high-level idea is that we apply randomized smoothing upon the inputs to a large language model and smooth its output, e.g., the decision on Drone's turning angles or flying directions. Specifically, our method divides a given input prompt into several masked prompts with disjoint subsets of tokens. Then, our method maps each token to an integer that indicates the index of the masked prompt. Then, our method assigns a token of the input prompt to the masked prompt. Then, our method predicts an output for each masked prompt, takes a majority vote based on an epsilon ball of each output, and then takes the averaged output as the final result. Since the method follows randomized smoothing, it will ensure that the output will not change much given an adversarial input.

In the past, our previous work has studied different attacks against LLMs. We will use our attacks to evaluate the robustness of the proposed GenAI system.

- Jailbreaking Attack. Our jailbreaking attack searches for alternative tokens in replacing the filtered tokens in a given prompt while still preserving the prompt's semantics and the follow-up generated images. Our high-level idea relies on Reinforcement Learning (RL), which adopts agents to interact with text-to-image models' outputs and change the next action based on rewards related to two conditions: (i) semantic similarity, and (ii) success in bypassing safety filters. Such RL agents not only solve the challenge of closed-box access to the text-to-image model but also minimize the number of queries as the reward function will guide the attack to find our adversarial prompts.
- Prompt Leaking Attack. Our novel, closed-box prompt leaking attack is inspired by existing jailbreaking attacks [17,22]. It optimizes a query, which we call adversarial query, such that a target LLM application is more likely to reveal its system prompt when taking the query as input. Specifically, we formulate finding such an adversarial query as an optimization problem, which involves a dataset of shadow system prompts and a shadow LLM. For each shadow system prompt, we simulate a shadow LLM application that uses the shadow system prompt and the shadow LLM. Roughly speaking, the objective of our optimization problem is to find an adversarial query, such that the shadow LLM applications output their shadow system prompts as the responses for the adversarial query.

5 Conclusions

Future autonomous systems need to have fail-safe conditions that are adaptive to dynamical and unpredicted conditions. We propose an architecture for autonomous attack recovery and outline how to make it more efficient and secure. Our future work will evaluate this architecture methodologically and in realistic conditions.

Acknowledgments. This material is based upon work supported in part by the Air Force Office of Scientific Research under award number FA9550-24-1-0015, and by the National Center for Transportation Cybersecurity and Resiliency (TraCR) (a U.S. Department of Transportation National University Transportation Center).

References

1. Chrysler recalls 1.4m vehicles for bug fix: https://www.wired.com/2015/07/jeep-hack-chrysler-recalls-1-4m-vehicles-bug-fix/, https://www.wired.com/2015/07/jeep-hack-chrysler-recalls-1-4m-vehicles-bug-fix/
2. Abdin, M., et al.: Phi-3 technical report: a highly capable language model locally on your phone. arXiv preprint arXiv:2404.14219 (2024)
3. Achiam, J., et al.: GPT-4 technical report. arXiv preprint arXiv:2303.08774 (2023)
4. Cárdenas, A.A., et al.: Attacks against process control systems: risk assessment, detection, and response. In: Proceedings of the 6th ACM symposium on information, computer and communications security, pp. 355–366 (2011)
5. Dash, P., Li, G., Chen, Z., Karimibiuki, M., Pattabiraman, K.: Pid-piper: Recovering robotic vehicles from physical attacks. In: 2021 51st Annual IEEE/IFIP International Conference on Dependable Systems and Networks (DSN), pp. 26–38. IEEE (2021)
6. Dayanıklı, G.Y., et al.: {Physical-Layer} attacks against pulse width {Modulation-Controlled} actuators. In: 31st USENIX Security Symposium (USENIX Security 22), pp. 953–970 (2022)
7. Dettmers, T., Lewis, M., Belkada, Y., Zettlemoyer, L.: GPT3. int8 (): 8-bit matrix multiplication for transformers at scale. Advances in Neural Inf. Process. Syst. **35**, 30318–30332 (2022)
8. Garg, K., Sanfelice, R.G., Cardenas, A.A.: Control barrier function-based attack-recovery with provable guarantees. In: 2022 IEEE 61st Conference on Decision and Control (CDC), pp. 4808–4813. IEEE (2022)
9. Gu, A., Dao, T.: Mamba: Linear-time sequence modeling with selective state spaces. arXiv preprint arXiv:2312.00752 (2023)
10. Jaffe, G., Erdbrink, T.: Iran says it downed us stealth drone; pentagon acknowledges aircraft downing. The Washington Post **5** (2011)
11. Liu, Z., et al.: MobileLLM: Optimizing sub-billion parameter language models for on-device use cases. arXiv preprint arXiv:2402.14905 (2024)
12. Malartic, Q., et al.: Falcon2-11b technical report. arXiv preprint arXiv:2407.14885 (2024)
13. Peng, B., et al.: Eagle and finch: RWKV with matrix-valued states and dynamic recurrence. arXiv preprint arXiv:2404.05892 (2024)
14. Rutkin, A.H.: spoofers use fake GPS signals to knock a yacht off course. MIT Technology Review (2013)
15. Sathaye, H., Strohmeier, M., Lenders, V., Ranganathan, A.: An experimental study of {GPS} spoofing and takeover attacks on {UAVs}. In: 31st USENIX Security Symposium (USENIX Security 22), pp. 3503–3520 (2022)
16. Shane, S., Sanger, D.E.: Drone crash in Iran reveals secret us surveillance effort. The New York Times **7** (2011)
17. Wallace, E., Feng, S., Kandpal, N., Gardner, M., Singh, S.: Universal adversarial triggers for attacking and analyzing NLP. arXiv preprint arXiv:1908.07125 (2019)
18. Xu, X., et al.: A survey on knowledge distillation of large language models. arXiv preprint arXiv:2402.13116 (2024)

19. Yan, C., et al.: Sok: a minimalist approach to formalizing analog sensor security. In: 2020 IEEE Symposium on Security and Privacy (SP), pp. 480–495 (2020)
20. Zhang, L., et al.: Fast attack recovery for stochastic cyber-physical systems. In: 2024 IEEE 30th Real-Time and Embedded Technology and Applications Symposium (RTAS), pp. 280–293. IEEE (2024)
21. Zhang, L., Chen, X., Kong, F., Cardenas, A.A.: Real-time attack-recovery for cyber-physical systems using linear approximations. In: Proceedings of the 2020 IEEE Real-Time Systems Symposium (RTSS), pp. 205–217. IEEE (2020)
22. Zou, A., Wang, Z., Kolter, J.Z., Fredrikson, M.: Universal and transferable adversarial attacks on aligned language models. arXiv preprint arXiv:2307.15043 (2023)

Open Access This chapter is licensed under the terms of the Creative Commons Attribution 4.0 International License (http://creativecommons.org/licenses/by/4.0/), which permits use, sharing, adaptation, distribution and reproduction in any medium or format, as long as you give appropriate credit to the original author(s) and the source, provide a link to the Creative Commons license and indicate if changes were made.

The images or other third party material in this chapter are included in the chapter's Creative Commons license, unless indicated otherwise in a credit line to the material. If material is not included in the chapter's Creative Commons license and your intended use is not permitted by statutory regulation or exceeds the permitted use, you will need to obtain permission directly from the copyright holder.

Autonomous Drone Swarms Using Lightweight LLMs

A. Azzouni[1] and G. Pujolle[2](✉)

[1] HeyCloud, Paris, France
[2] LIP6, Sorbonne University, Paris, France
Guy.Pujolle@lip6.fr

Abstract. This article explores the use of lightweight LLMs for controlling drone networks. We propose an approach leveraging lightweight LLMs capable of functioning in autonomous environments. The methodology involves piloting drone networks through voice-guided control and executing missions where vision processing plays a critical role. We also outline the chosen solution to secure communications, AI algorithms, and data. Finally, we present extensions planned for the next two years.

Keywords: Lightweight LLM · Drone swarms · Edge

1 Introduction

The rapid advancements in autonomous aerial robotics, particularly in drone swarms, have opened new opportunities for applications such as search and rescue, precision agriculture, and urban surveillance. However, controlling and coordinating a swarm to achieve complex, high-level objectives remains a significant challenge due to the need for robust and efficient decision-making algorithms that can operate on resource-constrained embedded platforms.

To address this, we propose a novel approach leveraging lightweight large language models (LLMs) running on edge devices to enable autonomous drone swarms to execute high-level commands through natural language interaction. Key components of the system include: 1) A speech-to-text conversion module to translate audio into text; 2) A natural language understanding module based on another LLM to parse the high-level objective and devise a step-by-step plan; 3) A language-to-code translation module to convert the plan into executable commands for the swarm; and 4) A feedback loop to continuously assess execution, refine plans, and adapt to the environment.

This approach builds upon recent advancements in the field of LLM-based autonomous agents, as highlighted in the survey by Wang et al. (2023) [1] and the work on using LLMs for robotics tasks by Wang et al. (2024) [2]. Furthermore, our work extends the multimodal human-agent interaction framework proposed by Nwankwo and Rueckert (2024) [3], leveraging the capabilities of pre-trained language and visual models to enable seamless natural language communication between humans and the drone swarm.

2 Background and Related Work

Existing approaches to controlling drone swarms often rely on complex algorithms and specialized hardware, making them inaccessible to non-expert users. Recent works have explored using natural language as a more intuitive interface for commanding robots, including field and service robots in unknown environments.

For instance, Walter et al. (2021) [4] demonstrated natural language control of robots in unstructured environments, while Nair et al. (2021) [5] proposed frameworks for learning language-conditioned behaviors from offline data. Similarly, Nwankwo and Rueckert (2024) [3] present a system for multimodal human-agent interaction using pre-trained language and visual models. Our proposed system builds on this line of work, leveraging the power of LLMs to enable autonomous drone swarms to execute high-level tasks through natural language.

Another relevant study by Bucker et al. (2022) [6] investigates using natural language commands to reshape robot trajectories, highlighting the potential of language-based interactions in robotics.

However, these approaches often demand significant computational resources, making deployment on edge devices challenging. Our work addresses this gap by employing lightweight LLMs that operate efficiently on resource-constrained platforms.

3 Proposed Approach

The proposed system consists of four key components:

1. **Speech-to-Text Conversion**: The user's high-level objective is captured through a microphone, and the audio is converted to text using a pre-trained LLM-based speech recognition model, such as OpenAI Whisper (Radford et al., 2022 [7]).
2. **Objective Understanding and Planning**: The text from the speech-to-text conversion module, along with the object annotations, is fed into a natural language understanding LLM, which generates a step-by-step plan to achieve the high-level objective.
3. **Object Detection and Annotation**: The drone's on-board camera streams are processed using a lightweight object detection model, such as YOLOv8 Nano, to identify relevant objects in the environment.
4. **Plan Execution and Feedback**: The generated plan is translated into executable commands for the drone swarm, which are transmitted and executed one by one. After each step, the system evaluates the outcome and updates the plan if necessary, creating a continuous feedback loop to ensure the successful completion of the objective.

As can be seen, our approach is mainly focused on using very lightweight architectures and models, enabling the entire system to be deployed on resource-constrained embedded platforms, such as the Nvidia Jetson Nano to enable autonomous drone swarms to execute high-level objectives. For example, we do not intent to use Lidar or any other expensive sensors, and instead rely on the onboard camera and lightweight object detectors, which adds additional challenges but also improves the cost-effectiveness and accessibility of our solution.

3.1 System Architecture

Our approach introduces innovations through edge-based lightweight LLMs and multimodal human-drone interaction.

3.1.1 Centralized vs. Distributed Architecture

We explore both centralized and distributed approaches to the LLM-based control system. The three architectures are shown in Fig. 1.

- **Centralized Approach**: A single, high-performance LLM processes all sensor data and generates plans. This approach offers a unified decision-making process but is limited by the computational resources available on the edge device.
- **Distributed Approach**: Each drone in the swarm is equipped with a lightweight LLM (or part of a larger LLM), allowing for scalable and resilient performance. Drones can collaborate to collectively understand the objective and generate a coordinated plan, leveraging their individual sensor data and processing capabilities.
- **Hybrid Approach**: A combination of the above, where a central LLM handles high-level planning and coordination, while distributed lightweight LLMs on each drone manage local perception and plan refinement.

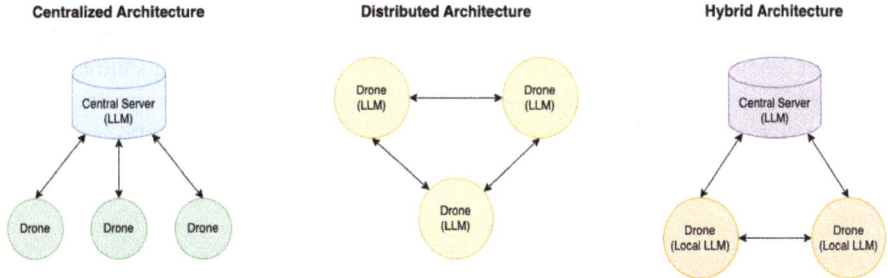

Fig. 1. Possible architectures

3.1.2 Comparison of Architectures

A comparative analysis of centralized, distributed, and hybrid architectural approaches reveals their distinct characteristics across four key criteria: failure tolerance, scalability, responsiveness, and computational efficiency.

In terms of failure tolerance, each architecture presents unique advantages and challenges. Centralized systems exhibit inherent vulnerability due to their single point of failure – the central server. While redundancy and backup systems can mitigate this risk, they introduce additional complexity to system management. In contrast, distributed architectures demonstrate robust failure tolerance, as the system remains operational even when individual nodes fail. The network can dynamically adapt by redistributing tasks among functioning nodes. Hybrid architectures strike a balance, offering partial independence through local nodes that continue operating during central system failures. However, they remain partially vulnerable when central coordination is essential for complex decision-making processes.

Regarding scalability, the architectures present distinct characteristics. Centralized systems face inherent limitations due to the bottleneck created by routing all data through a central server. As the system expands, the central server becomes increasingly strained, necessitating costly hardware upgrades or cloud scaling solutions. Distributed architectures excel in scalability by dispersing workloads across multiple nodes. This design enables efficient horizontal scaling through the addition of new nodes, with each node managing its local resources independently. Hybrid architectures similarly demonstrate strong scalability potential, combining local node processing with centralized oversight. This approach facilitates system growth through node addition while maintaining centralized control over global parameters.

In terms of responsiveness, each architecture offers different advantages. While centralized systems traditionally provide good response times, they can face challenges in real-time control scenarios requiring ultra-fast reactions. Distributed architectures achieve enhanced responsiveness through localized decision-making, positioning processing closer to data sources—a crucial advantage for real-time applications like autonomous vehicles and robotics. Hybrid architectures optimize responsiveness by enabling quick, local decisions for simple tasks while reserving complex processing for the central system, introducing latency only when necessary for more sophisticated operations.

Computational efficiency varies significantly across these architectures. Centralized systems often struggle with resource utilization, requiring substantial computing power to handle peak loads regardless of typical usage patterns. Distributed architectures maximize efficiency through localized processing and dynamic resource allocation across the network, with each node requiring only sufficient resources for its specific tasks. Hybrid architectures achieve balanced efficiency by relegating simpler tasks to local nodes while leveraging the central system for complex operations, creating an optimal distribution of computational resources.

This comparison is summarized in Table 1.

Table 2. Distributed Llama Performance Comparison

Model	1 x VM	2 x VM	4 x VM
Llama 2 7B	101.81 ms	69.69 ms	53.69 ms
Llama 2 13B	184.19 ms	115.38 ms	86.81 ms
Llama 2 70B	909.69 ms	501.38 ms	293.06 ms

4 Implementation

4.1 Key Components

To start, we opt for hybrid approach, with a central LLM for high-level planning and distributed lightweight LLMs on each drone for local perception and plan refinement. Here are the key components:

- **Speech-to-Text Conversion**: We use the OpenAI Whisper model, a state-of-the-art speech recognition LLM, to convert the user's voice commands into text. This model runs orchestrator (central computer).
- **Object Detection and Annotation**: We use the YOLOv8 Nano model, a lightweight object detection model, to identify relevant objects in the drone's camera feed. Each drone runs its own YOLO module locally.
- **Objective Understanding and Planning**: A pre-trained LLM, Llama-3.1-13B, is used to understand the user's high-level objective from the text input and generate a step-by-step plan to achieve it. This LLM runs on the central computer.
- **Plan Execution and Feedback**: The generated plan is executed by the drone swarm, with each drone controlled by a lightweight LLM for local perception and plan refinement.

The key components, and the hybrid architecture are shown in Fig. 2 and Fig. 3.

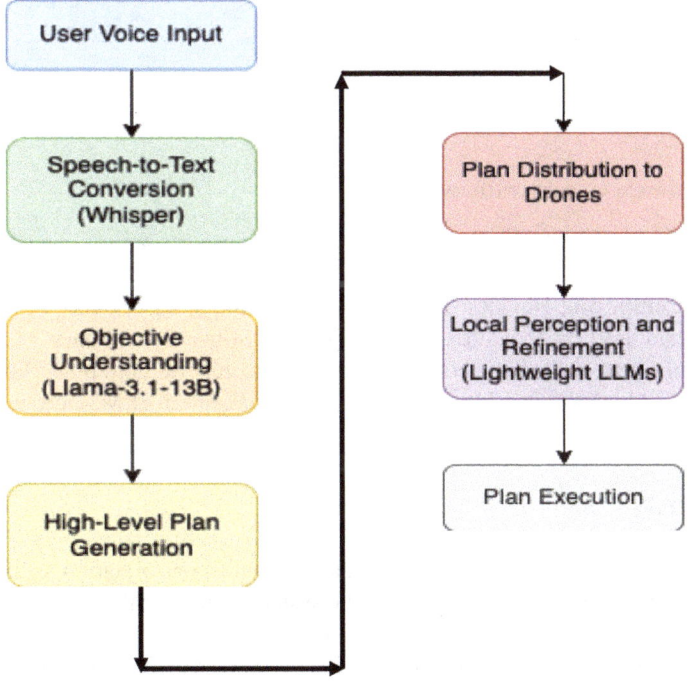

Fig. 2. The key implementation components

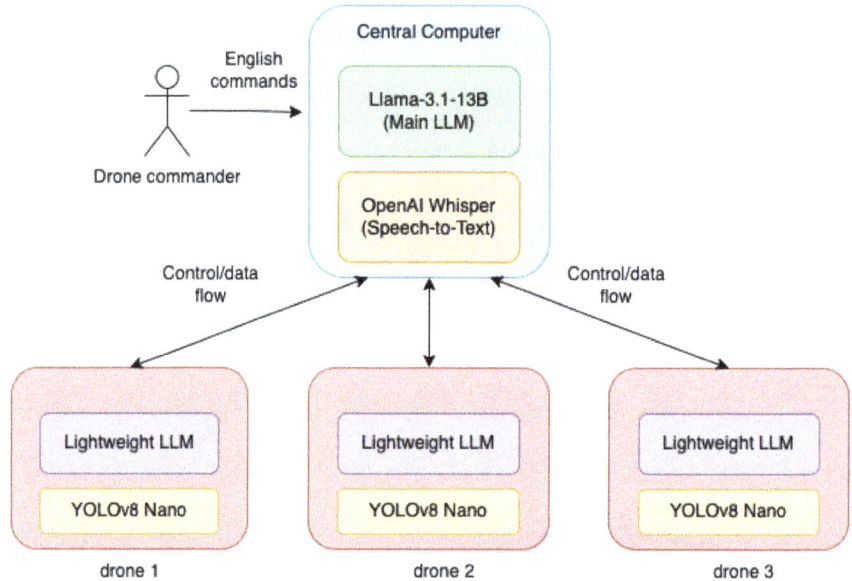

Fig. 3. The hybrid architecture

4.2 Hardware Considerations

- **Drone Platforms**: We use a swarm of lightweight, resource-constrained drones, based on the Pixhawk X500 platforms.
- **Edge Computing Devices**: Low-power edge computing devices, a Nvidia Jetson Nano card, host the LLM-based control system on-board the drones
- **Central Control Station**: For the hybrid architecture, a more powerful computing device, a Macbook Pro (i9 processor), hosts the central LLM and coordinates the swarm.

5 Security

Securing communication between drones and the central control station, while protecting AI algorithms and data, is crucial due to the system's vulnerability to potential attacks. To address this, we integrated EAL7-certified Hardware Security Modules (HSMs) into each drone and the central control station. These HSMs provide a tamper-resistant environment for the generation, storage, and management of cryptographic keys, ensuring system security and integrity.

HSMs serve as fundamental security components within the drone swarm and ground station infrastructure, primarily through their robust cryptographic key management capabilities. They generate, store, and protect the cryptographic keys essential for secure communications, enabling mutual authentication and encrypted data transmission.

The integration of HSMs extends to the implementation of Transport Layer Security (TLS) protocol, securing all inter-drone communications. Furthermore, HSMs digitally sign all commands and messages transmitted between drones and the ground station, ensuring message integrity. Any tampering during transmission becomes immediately detectable through signature verification failures.

In the event of drone capture, the HSM's sophisticated hardware security mechanisms prevent physical extraction of cryptographic keys. This protection ensures that adversaries cannot decode communications or compromise drone operations, maintaining system integrity even under physical security threats.

An HSM can also be used to sign drone software updates. This ensures that only authentic updates from an authorized source can be installed on the drone.

The Hardware Security Module (HSM) is a powerful tool for enhancing the security of AI systems by safeguarding critical elements of the AI pipeline. It provides advanced protection mechanisms to secure models, encrypt sensitive data, and ensure the integrity and authenticity of AI algorithms.

Specifically, the HSM serves as a robust security infrastructure for AI systems. For instance, model parameters can be encrypted using format-preserving encryption (FPE) based on the FF3-1 algorithm, enabling computations on encrypted data without compromising model accuracy. To protect models, the HSM implements secure model loading through authenticated encryption with AES-GCM-SIV for both weights and architecture. Additionally, runtime integrity verification is facilitated by continuously computing Merkle trees over model layers. Memory encryption is further enhanced using hardware-accelerated AES-XTS with 256-bit keys.

The HSM also strengthens data protection pipelines. It supports secure aggregation protocols for federated learning and enables the use of zero-knowledge proofs for verifying data provenance without revealing sensitive information. To ensure the integrity and authenticity of AI algorithms, methods such as Ed25519 signature-based code signing, secure boot chain verification for AI hardware, and version control with signed Git commits (secured by GPG keys stored in the HSM) can be employed.

Anti-reverse engineering measures are another critical capability of the HSM. These include obfuscating model architecture via virtual black box techniques and performing control flow integrity checks. Additionally, the HSM can prevent model theft through parameter watermarking and detect model extraction attempts via behavior monitoring.

This security architecture provides comprehensive protection for AI assets while maintaining operational efficiency. Regular security audits, penetration testing, and formal verification of critical security properties ensure robust defense against emerging threats. Automated security response mechanisms, including model rollback capabilities and secure recovery procedures, further enhance system resilience.

In summary, by managing critical cryptographic operations, the HSM significantly enhances the protection of AI models against threats such as theft, corruption, or reverse engineering, ensuring both security and reliability.

The hardware and software architecture of the HSM is shown in Fig. 4. The cost is less than US$ 100 and the weigh is less than 50 g [8].

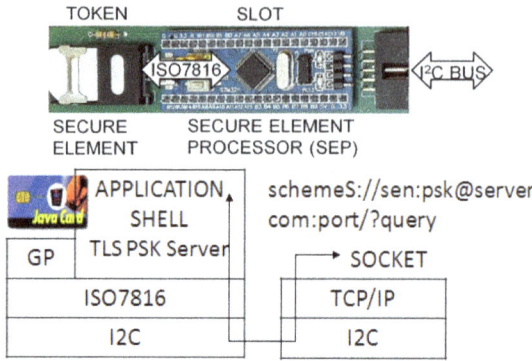

Fig. 4. Hardware and software architecture of the HSM

6 Experimental Evaluation

6.1 Simulation Experiments

We setup a high-fidelity simulation environment to test the performance of our system under various scenarios, including:

- Search and rescue operations in urban environments.
- Precision agriculture tasks, such as monitoring crop health and applying targeted fertilizers.
- Surveillance and inspection missions in hazardous areas.

The simulation results demonstrate the ability of our system to successfully execute high-level commands through the continuous feedback loop between the LLMs and the drone swarm.

6.2 Real-World Experiment

We also tested our system in a real-world setting using a fleet of two Nvidia Jetson-based Pixhawk drones equipped with the necessary hardware and software components. The real-world experiments focused on commanding the drones to explore an apartment looking for edible objects, such as oranges and other fruits. An example of communication between the drone and the operator of the swarm is described in Fig. 5.

7 Results and Discussion

Our experimental results indicate that the integration of lightweight LLMs on the edge enables autonomous swarms to execute high-level commands through natural language interaction with a high degree of success. The key advantages of our approach include enhanced adaptability, scalability, computational efficiency, and security.

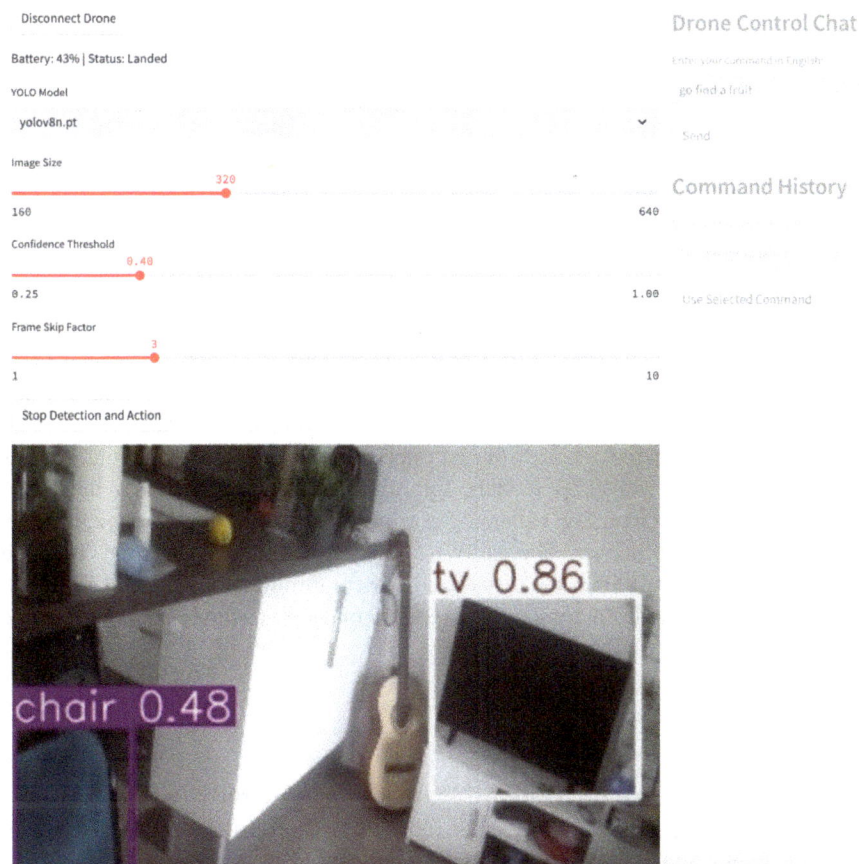

Fig. 5. The experiment

8 Future Work

The proposed system represents a significant advancement in enabling autonomous drone swarms for a broad range of applications. However, the current capabilities of LLM systems, particularly open-source vision LLMs like Llava (e.g., Llava:13b-v1.6-vicuna-q6_K), impose certain limitations.

That being said, the field is progressing rapidly, and we anticipate the emergence of more advanced vision LLMs within the next 12 to 18 months. A critical milestone will be achieving reliable object detection and distance inference with a 3B parameter vision LLM that operates with sub-one-second latency.

Table 1. Architecture comparison

Criteria	Centralized	Distributed	Hybrid
Failure Tolerance	Low	High	Moderate
Scalability	Low	High	High
Responsiveness	High	Moderate	High
Computational Efficiency	Low	High	Moderate

Another promising avenue we plan to pursue is distributing a larger LLM, including a vision LLM, across the drone swarm. For this, we are contributing to an early project called Distributed Llama (https://github.com/b4rtaz/distributed-llama), which divides LLM workloads among multiple devices to achieve substantial speed improvements. This system leverages TCP sockets for state synchronization via a home router.

We will also develop the use of HSMs to protect all components and communications of the drone swarm, as well as the models, algorithms, and data embedded in the drone.

The swarm will use GreenSoft technology from the company Green Communications for interconnecting the drones. This software integrates a blockchain used to define responsible applications within the distributed system through its consensus protocol. The software also includes a comprehensive suite of tools available across the drone swarm: video, storage, SDK, etc. [9].

9 Conclusion

In conclusion, over the next two years, we aim to develop a swarm of drones that is fully autonomous and capable of handling a wide range of missions using various SLMs with private data to ensure the reliability required for autonomy.

References

1. Wang, L., et al.: A Survey on Large Language Model Based Autonomous Agents. Cornell University (2023). https://doi.org/10.48550/arxiv.2308.11432
2. Wang, J., et al.: Large Language Models for Robotics: Opportunities, Challenges, and Perspectives. Cornell University (2024). https://doi.org/10.48550/arxiv.2401.04334
3. Nwankwo, L., Rueckert, E.: Multimodal Human-Autonomous Agents Interaction Using Pre-Trained Language and Visual Foundation Models. Cornell University (2024). https://doi.org/10.48550/arxiv.2403.12273
4. Walter, M.R., et al.: Language Understanding for Field and Service Robots in a Priori Unknown Environments. Cornell University (2021). https://doi.org/10.48550/arxiv.2105.10396
5. Nair, S., Mitchell, E., Chen, K., Ichter, B., Savarese, S., Finn, C.: Learning Language-Conditioned Robot Behavior from Offline Data and Crowd-Sourced Annotation. Cornell University (2021). https://doi.org/10.48550/arxiv.2109.01115
6. Bucker, A., Figueredo, L.F.C., Haddadin, S., Kapoor, A., Ma, S., Bonatti, R.: Reshaping Robot Trajectories Using Natural Language Commands: A Study of Multi-Modal Data Alignment Using Transformers. Cornell University (2022). https://doi.org/10.48550/arxiv.2203.13411

7. Radford, A., Kim, J.W., Xu, T., Brockman, G., McLeavey, C., Sutskever, I.: Robust Speech Recognition via Large-Scale Weak Supervision. arXiv preprint arXiv:2212.04356
8. Pujolle, G., Urien, P.: A New Generation of Security for the 6G, CSNet'20024, IEEE Xplore, Paris, France (2024)
9. Pujolle, G., Azzouni, A., Nguyen, T.M.T.: CloudNet'2024, IEEE Xplore, Rio de Janeiro, Brazil (2024)

Open Access This chapter is licensed under the terms of the Creative Commons Attribution 4.0 International License (http://creativecommons.org/licenses/by/4.0/), which permits use, sharing, adaptation, distribution and reproduction in any medium or format, as long as you give appropriate credit to the original author(s) and the source, provide a link to the Creative Commons license and indicate if changes were made.

The images or other third party material in this chapter are included in the chapter's Creative Commons license, unless indicated otherwise in a credit line to the material. If material is not included in the chapter's Creative Commons license and your intended use is not permitted by statutory regulation or exceeds the permitted use, you will need to obtain permission directly from the copyright holder.

Quantization-Based Privacy Preservation for Federated Learning in the Sky

Lamees M. Al Qassem[1,2,3(✉)], Maurizio Colombo[1], Ernesto Damiani[2,3,4], Rasool Asal[5], Al Anoud Almemari[1,2], and Yousof Alhammadi[3]

[1] EBTIC, Khalifa University, Abu Dhabi, UAE
{lamees.alqassem,maurizio.colombo,100041410}@ku.ac.ae
[2] Center for Cyber Physical Systems (C2PS), Khalifa University, Abu Dhabi, UAE
ernesto.damiani@ku.ac.ae
[3] Department of Computer Science, Khalifa University, Abu Dhabi, UAE
yousof.alhammadi@ku.ac.ae
[4] Computer Science Department, Università degli Studi di Milano, Milan, Italy
[5] College of Engineering and IT, University of Dubai, Academic City, Emirates Road, Dubai, UAE
rasool.asal@ud.ac.ae

Abstract. Unmanned aerial vehicles (UAV) have been widely used in various sectors, including military, emergency response, and space exploration. *Federated Learning in the Sky* (FLS) within UAV swarms is a new paradigm [11] which offers a cost-effective and efficient solution for data collection and analysis across a wide range of UAV applications. However, given the sensitive nature of the data exchanged within these swarms, they are highly vulnerable to eavesdropping and other cyber-attacks [1]. Our research targets secure and privacy-preserving communication of ML data and parameters within UAV swarms. We propose a lightweight communication framework for hierarchical drone swarms based on a randomized multi-hash data representation (*randomized hash comb* [6]) and on Feldman's secret sharing. The proposed solution is easy-to-implement, flexible, applicable to various architectures, and introduces minimal communication overhead. Our randomized hash comb algorithm guarantees Rényi differential privacy (RDP) of the training data and model updates exchanged among the UAVs. Feldman's protocol is used to securely share the initial parameters of the hash comb and to negotiate the hash values even in the presence of hijacked or rogue participants. Together, these techniques form a robust, secure, and efficient communication and coordination framework for data interchange within hierarchical UAV swarms.

Keywords: Quantization · Privacy · Distributed Learning · Federated Learning · Differential Privacy

1 Introduction

Drones are often deployed in swarms to process and collect data. For example, they can be integrated into Federated Learning in the Sky (FLS) systems [11], where model parameters are exchanged during missions. UAVs are increasingly utilized for collaborative missions and are becoming integral components of critical applications such as military recce and space exploration, which are targets of data breaches and attacks [1].

The growing popularity of drone swarms has raised concerns about the security and privacy of UAVs. In addition, there are concerns about data integrity due to the fact that drones not only transmit data but also store it locally or at the base station. Data Privacy threats have attracted the attention of privacy regulatory bodies worldwide [4]. For instance, the European General Data Protection Regulation (GDPR) imposes strict obligations on organizations anywhere so long as they target or collect data related to people in the EU. One of the key requirements impacting the data collection and processing by drones is the principle that GDPR-compliant applications should include data protection *by design and by default*. Therefore, there is a need to develop a comprehensive communication and coordination framework that ensures secure, privacy-preserving, and efficient operations for drone swarms, especially hierarchical ones. However, the main challenge is developing a robust, lightweight framework that can work reliably in a dynamic environment under adversarial conditions.

Collaborative development of machine learning (ML) models often happens in adversarial settings where participants may have conflicting interests and varying levels of trustworthiness. In this context, UAV swarms from different organizations may dynamically change their level of collaboration in each learning round, making it difficult to guarantee consistent contributions from all participants. This dynamic environment is vulnerable to different types of attacks, such as data poisoning attacks, where malicious drones may inject false data to bias the global model in their favor; Byzantine attacks, where compromised UAVs send corrupted model updates; and eavesdropping and data leakage, where untrusted drones interfere or intercept sensitive information from the shared model updates [3]. Therefore, a verifiable mechanism is crucial to maintain data integrity and privacy, especially in critical applications like landmine detection [8] and traffic prediction [10].

In this context, our research problem is designing a communication and coordination framework that supports privacy protection, suitable for being deployed in hierarchical drone swarms composed of diverse UAVs having different on-board resources. Our proposed architecture, depicted in Fig. 1, consists of edge devices responsible for collecting and transmitting data, while higher-level aggregators (i.e., fog drones) handle data aggregation, processing, and analysis. In this paper, we present our communication platform and discuss the key findings from its implementation in a distributed learning system.

2 Scenario Description

Our communication platform designed for hierarchical drone swarms is shown in Fig. 1. Edge devices collect data and transmit it to fog drones, which serve as clients in the Federated Learning in the Sky (FLS) model. These data transmissions may involve personal and sensitive information (for instance, images of people or buildings), necessitating privacy-preserving measures. Prior research [9] indicates that private information can be extracted from ML model parameters when they are trained directly on raw data; so, a privacy-preserving data transformation must be applied. Nevertheless, the data must be kept suitable for updating on-board ML models as part of the FLS system.

To address these challenges, our platform incorporates data protection techniques across multiple levels: (1) from the perspective of data owners (i.e., edge devices) whose data is transmitted and processed by fog drones, and (2) from the perspective of FL clients (i.e., fog drones) that manage training datasets, machine learning model parameters, or both, and seek to safeguard this information during training, irrespective of the specific machine learning architectures or algorithms employed. Our goal is to maintain the confidentiality of local model parameters while enabling the accurate computation of a global model.

Furthermore, it is important to note that drones may have limited computational resources. Our communication platform is designed with this limitation in mind, especially in scenarios where encoding must be performed in a CPU-only environment. This consideration is critical for ensuring the feasibility and efficiency of the system under constrained computational conditions.

To tackle these challenges, we achieve Rényi differential privacy of the transmitted data by applying a randomized quantization. We use the *Randomized Hash-Comb* technique, a novel and lightweight multi-hash data representation method introduced in [6] and built upon the multi-quantization encoding framework proposed in [2].

However, our randomized quantization technique necessitates secure sharing across the swarm of the quantization process hyper-parameters. The secret sharing must work even in the presence of several hijacked or rogue drones that try to inject forged or faulty values. We tackle this problem by integrating Feldman's secret sharing protocol [5], which enables verifiable, secure secret sharing even in the presence of a dishonest majority of participants. To the best of our knowledg,e this is the first time that robust secret sharing is proposed for drone swarms. Integrating Feldman's secret sharing strengthens our framework's robustness and ensures data integrity across all levels of the drone swarm, even in the presence of untrusted UAVs.

3 Key Findings

Our communication platform has been tested on centralized federated learning architecture with multiple participants [6]. The experimental results show that the solution achieved the desired level of differential privacy (DP) of ML models' parameters and data exchanged in distributed training. This preserves the confidentiality of the model's parameters while maintaining the performance and accuracy of the global aggregated model. Moreover, the results show that the performance was intact with the dropping or random selection of the participants. This indicates that the randomized hash comb algorithm effectively adapts to dynamic conditions, maintaining secure communication channels and efficient

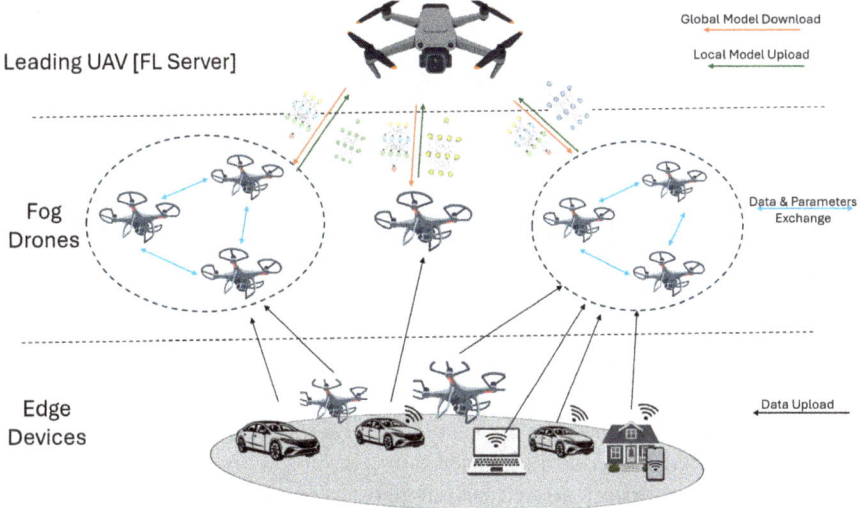

Fig. 1. Hierarchical Drone Architecture where the edge devices transfer data to fog drones to process and perform local model training. These local models are sent to the leading UAV for aggregation and updating the global model of the federated learning system.

task completion. Furthermore, these outcomes are consistent with our hierarchical architecture, where the training process in distributed learning (DL) is conducted across multiple machines or nodes with varying computational power.

By adopting a robust, secure communication scheme like Feldman secret sharing [5], we can securely add participants (i.e., drones) even in the presence of a dishonest majority. This ensures the integrity and reliability of communications under adversarial conditions. The Feldman protocol adds an additional layer of security that maintains the system's integrity.

Our technique relies on standard SHA-3 hashes, easy-to-compute via FPGAs even when computational resources are limited [7]. It also leverages quantization to reduce the communication overhead, which is a known bottleneck in hierarchical systems.

Finally, our system is designed with a hierarchical architecture in mind, where edge devices handle initial data processing and fusion, while fog drones, with their higher computational capacity, aggregate data from multiple edge devices and perform swarm-level analysis. By integrating this hierarchical data fusion with our proposed secure communication framework, we ensure that data integrity is preserved at all levels within the drone swarm, enabling more robust threat detection and response capabilities. Additionally, this approach aligns with regulatory authorities' objectives by facilitating the free movement of data, provided that adequate safeguards are in place.

4 Conclusion

Progress in the technology underlying large UAV swarms has enabled promising UAV-based computation paradigms like Federated Learning in the Sky. On the other hand, the growing complexity of drone swarms makes them vulnerable to cyber-attacks. Our proposed communication framework leverages the randomized hash-comb data representation to quantize the data or parameters shared in distributed learning protocols and uses Feldman's algorithm to securely negotiate the quantization parameters. By combining randomized data quantization with a secure negotiation protocol, we aim to significantly enhance the security, privacy, and efficiency of communication operations in drone swarms.

In summary, we claim that our technique's ability to measurably improve privacy, enable data utility, and adapt to the requirements of different industries make it a powerful framework for deploying hierarchical drone swarms even in highly regulated domains like cities' surveillance. Our approach effectively addresses some important security challenges to applications based on drone swarms, paving the way to achieving compliance to data protection and privacy regulations.

References

1. Albalawi, M., Song, H.: Data security and privacy issues in swarms of drones. In: 2019 Integrated Communications, Navigation and Surveillance Conference (ICNS), pp. 1–11. IEEE (2019)
2. Almahmoud, A., Damiani, E., Otrok, H.: Hash-comb: a hierarchical distance-preserving multi-hash data representation for collaborative analytics. IEEE Access **10**, 34393–34403 (2022)
3. Alsamhi, S.H., et al.: Computing in the sky: a survey on intelligent ubiquitous computing for uav-assisted 6g networks and industry 4.0/5.0. Drones **6**(7) (2022). https://doi.org/10.3390/drones6070177, https://www.mdpi.com/2504-446X/6/7/177
4. Caroline, B., et al.: Artificial intelligence cybersecurity challenges; threat landscape for artificial intelligence (2020)
5. Chen, Y.H., Lindell, Y.: Feldman's verifiable secret sharing for a dishonest majority. Cryptology ePrint Archive (2024)
6. Colombo, M., et al.: A quantization-based technique for privacy preserving distributed learning (2024). https://arxiv.org/abs/2406.19418
7. Kaps, J.P., et al.: Lightweight implementations of SHA-3 candidates on FPGAs. In: Progress in Cryptology–INDOCRYPT 2011: 12th International Conference on Cryptology in India, Chennai, India, December 11-14, 2011. Proceedings 12, pp. 270–289. Springer (2011)
8. Kasianchuk, A., Lastivka, H.: UAV integration with neural network in landmine and minefield detection tasks. Security of Infocommunication Systems and Internet of Things **1**(2), 02008–02008 (2023)
9. Lang, N., Sofer, E., Shaked, T., Shlezinger, N.: Joint privacy enhancement and quantization in federated learning. IEEE Trans. Signal Process. **71**, 295–310 (2023)

10. Lim, W.Y.B.: Towards federated learning in UAV-enabled internet of vehicles: a multi-dimensional contract-matching approach. IEEE Trans. Intell. Transp. Syst. **22**(8), 5140–5154 (2021)
11. Liu, Y.: Federated learning in the sky: Aerial-ground air quality sensing framework with UAV swarms. IEEE Internet Things J. **8**(12), 9827–9837 (2020)

Open Access This chapter is licensed under the terms of the Creative Commons Attribution 4.0 International License (http://creativecommons.org/licenses/by/4.0/), which permits use, sharing, adaptation, distribution and reproduction in any medium or format, as long as you give appropriate credit to the original author(s) and the source, provide a link to the Creative Commons license and indicate if changes were made.

The images or other third party material in this chapter are included in the chapter's Creative Commons license, unless indicated otherwise in a credit line to the material. If material is not included in the chapter's Creative Commons license and your intended use is not permitted by statutory regulation or exceeds the permitted use, you will need to obtain permission directly from the copyright holder.

Human-Drone Swarm Collaboration Using LLMs: Case Study on DRL-Based Anti-jamming

Abubakar S. Ali[1(✉)], Shimaa Naser[1], Omar Alhussein[1], Sami Muhaidat[1,2], and Ernesto Damiani[1]

[1] Khalifa University, Abu Dhabi, UAE
asali.ele@buk.edu.ng , shimaa.naser@ku.ac.ae
[2] Carleton University, Ottawa, Canada
https://www.ku.ac.ae

Abstract. As autonomous systems such as drone swarms become increasingly crucial in complex missions, it is essential to ensure effective human oversight and explainable human-machine collaboration. We propose to integrate large language models (LLMs) as an interface with artificial intelligence (AI) agents to enhance explainability and incorporate human-in-the-loop control. We discuss how LLMs can bridge the gap between the technical complexities of AI-based autonomous systems and post-hoc interpretation techniques. LLM as an interface can enhance human-machine collaboration and control and increase trust and safety. Moreover, it has the potential to provide emergent capabilities and enhanced meta-learning. In our preliminary work, we integrate LLMs with deep reinforcement learning (DRL) to enhance anti-jamming capabilities of autonomous systems. Through a detailed case study, we demonstrate how this approach not only mitigates jamming threats but also facilitates human-in-the-loop control, enabling dynamic adjustments to mission parameters. Additionally, the natural language interface provided by LLMs enhances communication efficiency between human operators and drone swarms, ensuring seamless collaboration and improved operational resilience.

Keywords: 6G · anti-jam · deep reinforcement learning · explainable AI · large language models

1 Introduction

The evolution of modern networks has introduced a new era of communication and automation. As we transition to sixth-generation (6G) networking, we stand on the edge of unlocking significant advancements that will redefine how autonomous systems operate in complex environments. At the heart of this transformation is the integration of advanced artificial intelligence (AI) techniques, such as deep reinforcement learning (DRL) and large language models (LLMs), into autonomous systems such as drone and robot swarms.

Deep learning is essential for autonomous systems, enabling adaptation to dynamic environments with minimal human intervention. DRL, for example, helps counteract threats like jamming attacks on communication systems. However, the complexity and opacity of these models hinder human understanding and limit effective human-machine collaboration, especially in safety-critical applications where trust is vital [1]. To improve transparency, researchers have explored methods such as attention mechanisms, visualization tools, and explainable AI (XAI) approaches [2,3].

LLMs have the potential to bridge the gap between interpretable AI techniques, which typically require substantial technical expertise and offline analysis, and explainable AI that produces human-readable outputs. LLMs are not only capable of understanding and generating human language but also of interpreting and explaining the actions of AI agents in real-time. This capability is particularly valuable in scenarios where human operators (who are not necessarily engineers/analysts) need to interact with and oversee the operations of autonomous systems in real-time.

2 Proposed Framework

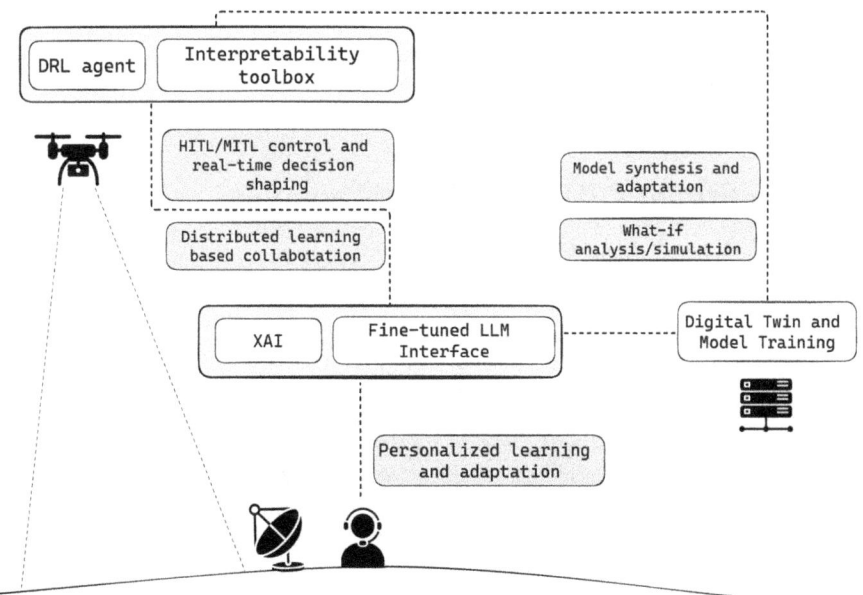

Fig. 1. Proposed framework for AI agents assisted with fine-tuned LLM-based interface.

2.1 LLM Interface and Human-in-the-Loop Control

A key feature of the proposed framework is an LLM-based natural language interface, which simplifies interactions between human operators and the AI agent. This interface enables operators to influence autonomous system decisions more easily and reduces response times in critical situations. For example, in an anti-jamming scenario, an operator can quickly ask about the risks of switching to a specific channel. The LLM interface adapts to individual operators' language and decision styles and supports continuous improvement through digital twins and LLM-assisted model synthesis and training [4]. Operators can interact with the system to get explanations and enact dynamic decision shaping. This fosters trust, enhances understanding, and allows operators to make informed decisions and recommendations.

2.2 Explainable AI

The integration of LLMs with deep learning systems plays a crucial role in enhancing the transparency of autonomous operations. At the heart of this integration is the transformative capability of the LLM to act as an explainable module. While various interpretability solutions exist, they are often complex to understand and require deep technical expertise. Moreover, there is no one-size-fits-all technique that addresses all scenarios and data distributions. In this context, LLMs can bridge the gap between interpretability and explainability. Distributed learning can be leveraged to design a distributed LLM-based explainable solution that is natively integrated with an AI agent. Multimodal language models present a promising direction towards further alleviating this performance-interpretability trade-off.

3 Case Study: DRL-Based Anti-jamming

3.1 DRL-Based Anti-jamming Strategy

Modern communication systems are vulnerable to jamming attacks, which can disrupt their operation. To counter this, we employ DRL to autonomously learn and implement anti-jamming strategies. Additionally, we integrate LLM to improve the explainability of the DRL agent's actions.

MDP Formulation and Agent Design:

In our scenario, the state space is a vector representing received jamming power across channels, while the action space is the choice of a communication channel. The agent receives rewards for avoiding jamming and minimizing channel-switching costs. We use a Double Deep Q-Network (DDQN) to reduce Q-value overestimation found in standard DQNs. The DDQN improves learning stability by using separate networks for estimating current and target Q-values. The agent's experience replay buffer stores transitions for training the Q-network via stochastic gradient descent. In the context of drone swarms, this approach allows autonomous systems to adapt to dynamic jamming conditions, ensuring reliable communication even in hostile environments. The agent's ability to learn and apply these strategies autonomously is crucial for maintaining the operational integrity of the swarm.

3.2 Integration of LLM for Enhanced Transparency

The LLM serves as an interpreter, translating complex DRL strategies into human-readable narratives. For example, after the DRL agent selects a channel based on its learned policy, the LLM generates an explanation that describes the reasoning behind this choice, e.g. "the agent selected channel 2 due to its low received jamming power, aiming for optimal throughput with minimal channel switching." This capability not only makes the system's actions understandable to human operators but also facilitates better decision-making and trust in the system's autonomous operations (Fig. 3).

3.3 Results and Analysis

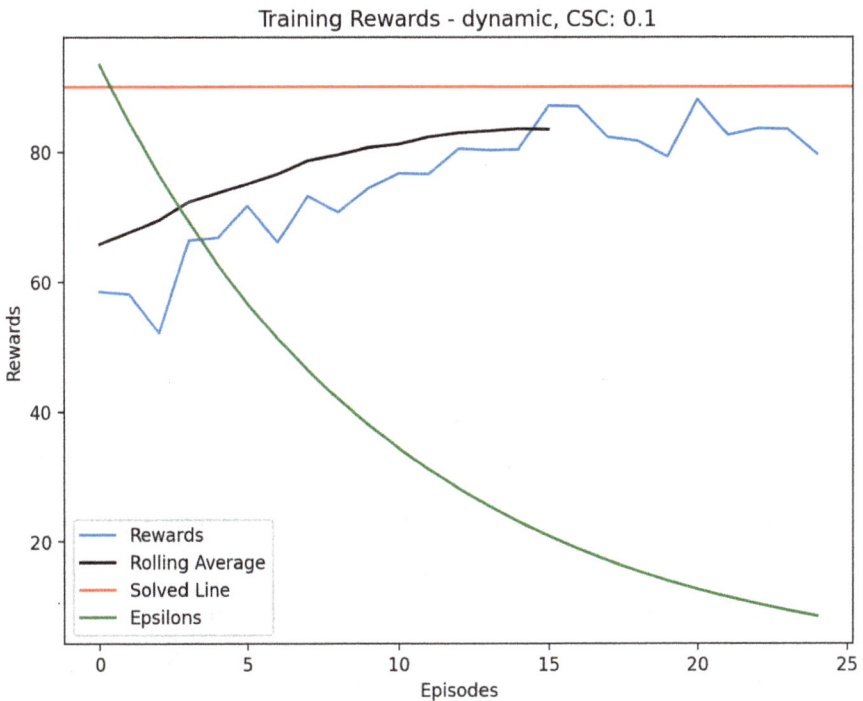

Fig. 2. Training performance of the DRL-based anti-jamming.

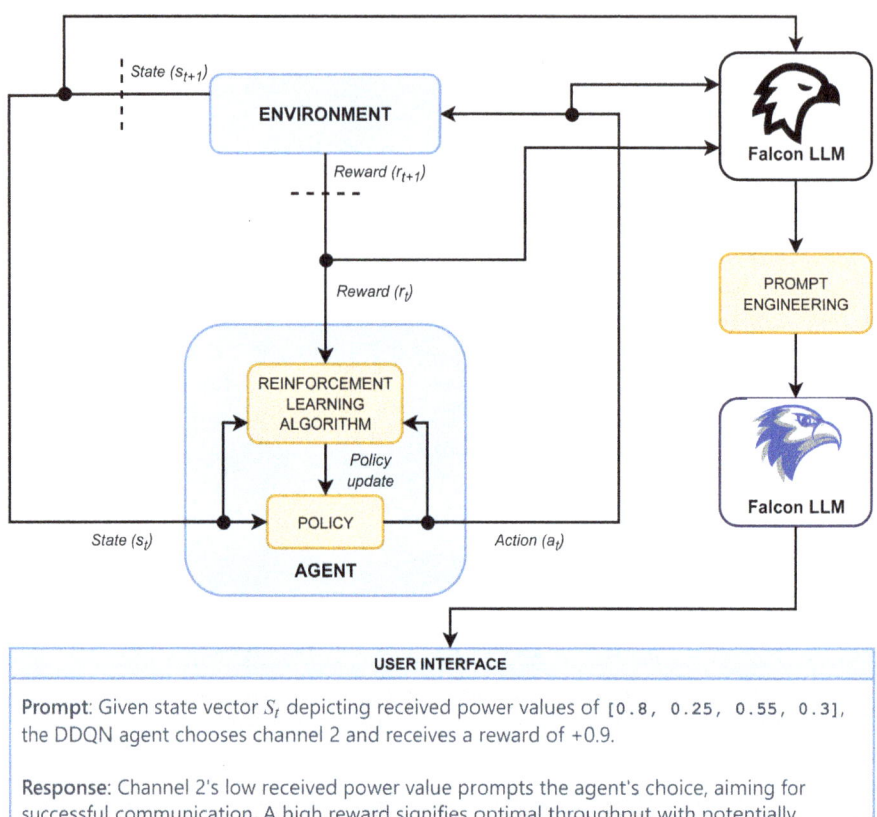

Fig. 3. Integration schematics of the DRL-based anti-jamming technique combined with LLM interpretation.

To evaluate the effectiveness of the proposed solution, we conducted a series of experiments where the DRL-based anti-jamming agent was trained and tested in various jamming scenarios. Figure 2 illustrates the training performance of the DRL-based agent, showing how the cumulative reward improves over time as the agent learns to avoid jamming. The LLM was used to interpret these results, providing human-readable insights into the agent's learning process, such as the effectiveness of the agent's strategy in different phases of training.

Within this realm, the integration of the 'falcon 7B' LLM amplifies the capacity for insightful data interpretation. Figure 1 demonstrates the schematic that elucidates how data from the DRL-agent training is ingested and processed. In our codebase, (https://huggingface.co/spaces/asataura/jam_shield_LLM_app) we use prompt engineering in the 'train' function within our 'trainer' module to showcase this. Here data about rewards, rolling averages, and epsilon values are fed to the LLM, to facilitate a richer understanding and optimizing network decisions.

3.4 User Interface Implementation on Hugging Face

In addition to the underlying DRL and LLM integration, we developed a user-friendly interface (UI) hosted on Hugging Face, which can be accessed at https://huggingface.co/spaces/asataura/jam_shield_LLM_app. This interface provides an accessible way for users to interact with and visualize the DRL-based anti-jamming system, making it easier to configure, monitor, and interpret the agent's performance. The UI allows users to:

Configure the Environment: Users can choose different jammer types (e.g., sweeping, random) and set parameters like channel switching cost, allowing for robust testing under various jamming scenarios.

Monitor DRL Training Progress: The UI offers real-time updates on DRL training, showing metrics such as episode count, rewards, and exploration rate, helping users track the agent's learning progress.

Generate Insights with LLM: After training, the LLM generates detailed insights from the data, interpreting graphs and providing human-readable explanations of the agent's behavior and achievements. This feature enhances transparency and ensures operators understand the rationale behind the actions taken.

4 Conclusions and Future Research Directions

We propose enhancing human-swarm collaboration in autonomous systems by integrating LLMs, DRL, and advanced explainability techniques. This approach aims to improve transparency, trust, human-in-the-loop control, and provide a natural language interface for greater reliability in complex environments. Moreover, digital twin technology with model synthesis environments can be leveraged to continually enhance DRL agents through LLM-assisted training with a human in the loop. In a recent work, we showcased how LLM can be leveraged to autonomously create curricula for DRL agents where generalization and convergence is improved [4]. This is an emerging field, open to innovative approaches for automating model training and evolution.

Our proposed framework far extends beyond anti-jamming to span surveillance and monitoring applications, autonomous navigation, human-robot collaboration, and LLM-enhanced native-AI communications. Future work can explore the integration of advanced interpretatability techniques with LLM-assisted active and explainable AI. In a recent work, we showcased how Bayesian deep learning can be leveraged to enhance the efficiency of data sampling and increase the interpretability of complex decision making [5,6]. Further research is needed in this direction.

LLMs should be optimized through fine-tuning and retrieval-augmented generation, and personalized for each operator and use case using in-context learning. In critical scenarios, grounding and model alignment are essential to avoid hallucinations and biased decisions from data leakage. In recent work, we have outperformed state-of-the-art LLMs like GPT-4 by developing a fine-tuned RAG-based medium-sized language model specialized in telecommunication networks [7]. Our team possesses the expertise and capabilities to deliver advanced, out-of-the-box, and practical human-machine LLM-assisted interface and control.

References

1. Du, Y., et al.: Guiding Pretraining in Reinforcement Learning with Large Language Models. arXiv, vol. abs/2302.06692 (2023)
2. Chen, W., et al.: Deep reinforcement learning for Internet of Things: a comprehensive survey. IEEE Commun. Surv. Tutor. **23**(3), 1659–1692. IEEE (2021)
3. Rjoub, G., Bentahar, J., Wahab, O.A.: Explainable AI-based federated deep reinforcement learning for trusted autonomous driving. In: Proceedings of IEEE IWCMC, pp. 318–323. IEEE (2022)
4. Erak, O., et al.: Large Language Model-driven Curriculum Design for Mobile Networks. CoRR, vol. abs/2405.18039 (2024). Accepted to Proceedings of IEEE/CIC ICCC
5. Alhussein, O., Akhavain, A.: Method and apparatus for managing network traffic via uncertainty. US20230216811A1, United States Patent and Trademark Office (2023)
6. Alhussein, O., Zhang, N., Muhaidat, S., Zhuang, W.: Active ML for 6G: Towards Efficient Data Generation, Acquisition, and Annotation. CoRR, vol. abs/2406.03630v1 (2024)
7. Alabbasi, N., et al.: An Oracle for Telecom Networks: Leveraging Fine-tuned Retrieval-Augmented Generation with Long-Context Support. To be submitted to IEEE Trans. Netw. Serv, Manag (2024)

Open Access This chapter is licensed under the terms of the Creative Commons Attribution 4.0 International License (http://creativecommons.org/licenses/by/4.0/), which permits use, sharing, adaptation, distribution and reproduction in any medium or format, as long as you give appropriate credit to the original author(s) and the source, provide a link to the Creative Commons license and indicate if changes were made.

The images or other third party material in this chapter are included in the chapter's Creative Commons license, unless indicated otherwise in a credit line to the material. If material is not included in the chapter's Creative Commons license and your intended use is not permitted by statutory regulation or exceeds the permitted use, you will need to obtain permission directly from the copyright holder.

Challenge 2: – Swarm-Level Threat Intelligence and Response System

Multi-modal Swarm Intelligence for Secure UAV Missions

Yunming Xiao(✉), Mushtari Sadia, and Ang Chen

University of Michigan, Ann Arbor, MI 48109, USA
yunmingx@umich.edu

Abstract. Unmanned Aerial Vehicles (UAVs) have found wide use in various tasks, but developing swarm intelligence for intrusion detection is far from easy. In this project, we propose to leverage recent advances in large multimodal models (LMMs) that can fuse multiple data sources for secure missions. Our project will fine-tune and deploy LMMs of varying sizes to the edge/fog UAVs, combining data sources such as sensory inputs (e.g., camera, IMU) as well as internal operational data (e.g., syscall logs). This will enable real-time detection and response system to thwart threats with swarm-wide coordination, addressing Challenge #2 outlined by the GENZERO workshop. Moreover, we aim to perform hardware-in-the-loop tests, with real devices, data, and scenarios, leveraging University of Michigan's M-Air 10,000 sq ft, four-story, testing facility. We envision that the final outcome will be an integrated system and demo validated in M-Air, achieving Technology Readiness Level 4 (TRL4).

Keywords: UAV Security · Swarm Intelligence · Large Multi-modal Model · Model Compression · Inference Scheduling

1 Introduction

Unmanned Aerial Vehicles (UAVs) have been widely deployed for various tasks, including aerial photography, agricultural monitoring, infrastructure inspection, emergency services, and smart cities [16]. A recent trend is to deploy UAV swarms in a hierarchical structure for a wider range of tasks. A canonical scenario includes (i) a fog UAV, which has relatively higher computational power, working alongside a swarm of (ii) edge UAVs; in addition, (iii) a command unit (e.g., a specialized vehicle) may be positioned nearby to better support the UAV swarm operation. UAV swarms, therefore, significantly extend our ability for critical missions.

However, the growth of the UAV market also attracts increased attention from attackers. Recent years have seen numerous security breaches in both military [1] and civilian [10] UAV operations. UAV vulnerabilities span communication channels, software, sensors, and supply chains [19]. Software vulnerabilities from traditional deployments can appear in UAVs, and their exposure to the environment invites physical attacks (e.g., GPS spamming, signal jamming). In swarms, the larger number of UAVs and sophisticated missions expand the attack surface. We are in dire need to develop a *zero-trust architecture for secure and resilient UAV missions*.

Existing intrusion detection projects do not fully exploit the unique nature of UAV swarms. First, defense mechanisms are implemented for each UAV and for each kind of threat individually in gps-detect-ml-24, uav-ids-traffic-22, uva-security-review-23. This raises the complexity of security management and misses important correlation analyses across a swarm for better, coordinated responses. Second, existing defense mechanisms have not sufficiently considered the rich data sources of individual UAVs – not only its sensory inputs (e.g., camera, IMU) that describe the external, physical environments, but also various software logs that capture the details of the internal operation. Hence, existing UAV intrusion detection systems only provide a siloed view of the security state, but they lack *swarm intelligence* that will be crucial for zero-trust UAV missions – that is, real-time, intelligent response to detected threats so no swarm component is granted automatic trust.

To address the above challenges, we propose a zero-trust architecture that is imbued with swarm intelligence. This architecture is made possible by two components. First, we propose a hierarchy intrusion detection system where each layer handles different problems (Fig. 1). Furthermore, the intrusion detection is powered by recent development of multimodal AI. Together, our proposed IDS correlates UAV system logs, sensory inputs, and network communication across the swarm to drastically improve the security of the mission.

Hierarchical intrusion defense: At the edge, fog, and command unit, our architecture deploys different versions of a pre-trained Large Multimodal Model (LMM), taking into account the varying computational and other constraints of each location. An interesting tradeoff is between model quality/size and inference latency. From the edge to the fog and to the command unit, the computational resources allow for larger LMMs which provide better inference quality and coordination capability; however, meanwhile, the model inference latency increases due to data transfer across long-distance communication. We will design our IDS with a scheduler that decides which model to query, for the overall goal of improving attack detection accuracy and timeliness.

Multimodal AI: Our observation is that UAVs are cyber-physical systems that produce rich sources of data about their operations. UAV software logs enable the IDS to gain deep understanding of the cyber aspects, whereas the sensory inputs provide insight into the physical aspects. By coupling these data sources together, we can perform better intrusion defense. The recent development of multimodal AI is a key enabler, as software logs, sensory inputs, and network communication have drastically different modalities (see subsect. 2.2).

Research roadmap: To achieve swarm intelligence, our proposed roadmap and project timeline are as follows.

- Training (one year): We will use the M-Air Test facility [3] to collect real-world data from UAVs (e.g., system and sensory logs, communication). We will also augment this data with system logs from non-UAV devices (e.g., datacenter servers) because UAVs face similar cyber threats (e.g., buffer overflow, malware [6]) as traditional computers due to shared software and OSes (e.g., Linux-based OSes [6]); hence, LMMs can benefit from these logs. We will fine-tune an open-source LMM with the additional data.
- Compression (six months): We will then successively compress the LMM so that they can be customized for constrained environments (e.g., at the edge/fog).
- Inference (six months): At the inference time, a scheduler will execute in each UAV device, deciding what data to collect and where the inference should occur. We will also explore parallelizing inference requests both locally and remotely, with deadlines set for remote completions.

2 System Design

2.1 Overview

As illustrated in Fig. 1, we propose a hierarchical UAV defense system aligned with the physical hierarchy: edge UAVs executing missions, a fog UAV coordinating them, and a remote server or specialized vehicle as the command unit. Our core idea is to use multimodal data – sensor data, system logs, and communication packets – to build a holistic defense mechanism against all threats, targeting Challenge #2 outlined in the GENZERO workshop.

Given the physical constraints, each level of the system hierarchy has different computational capabilities. For instance, edge UAVs may only have weak CPUs; fog UAVs might have GPUs; whereas the command unit, with much greater computational power, can support a fully-fledged LMM. Thus, for fog and edge UAVs, we compress the model in a task-specific manner. The compressed model satisfies local security defense demands, and when necessary, the lower hierarchy unit can query the upper hierarchy unit for more accurate and informative decisions against threats.

We categorize the procedure to realize our proposed system into two steps: training and inference, as detailed below.

2.2 Training

Training Data Sources. To train a robust LMM for secure UAV operations, high-quality data is essential. We will focus on UAV threats during the operational phase at both individual and swarm levels. Individual UAVs need to mitigate communication vulnerabilities, sensor manipulations [24], and system vulnerabilities like malware exploits [21], while UAV swarms must address intercommunication threats [4]. Our key research task is curating diverse training data to develop an LMM-based intrusion detection model for UAV swarms.

Our insight is that UAVs share many similarities with non-UAV scenarios, despite their unique characteristics [22]. From the software angle, UAVs face similar threats due to the adoption of shared open-source libraries and operating systems [6]. We thus plan to collect data from both real-world UAV tests and public datasets, as well as non-UAV scenarios like datacenter logs [2]. By integrating these diverse data sources, we can enhance our model's ability to detect and mitigate threats, leveraging techniques from established fields, such as datacenter SIEM tools [12], tailored to UAV challenges.

Finally, our training data also includes various sensor data, such as GNSS signals, IMU records, and camera footage. We plan to leverage both public datasets and data generated from UAV simulators and real-world operations. Additionally, we have noticed that satellite images of the operational region can enhance UAV resilience against sensory manipulation [14]. We consider incorporating these images as input as well.

Furthermore, we will explore GANs/LLMs [20] to create synthetic datasets that can simulate various threat scenarios and rare events. This synthetic data will augment our training datasets, providing additional scenarios and enhancing the model's robustness and generalization capabilities.

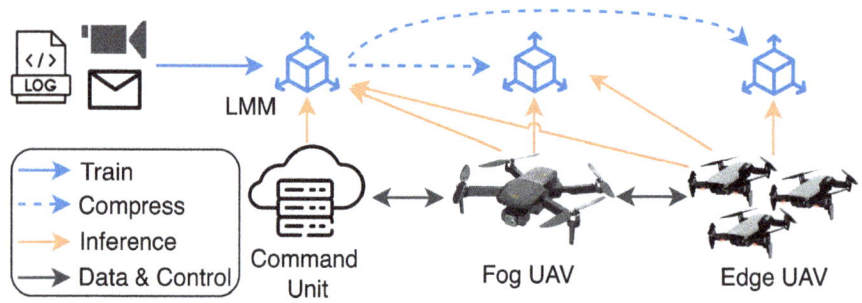

Fig. 1. Overview of system architecture.

Model Training. With the collected data, we will fine-tune existing LMMs rather than training a new one from scratch because (*i*) state-of-the-art LMMs already incorporate extensive and valuable data for effective reasoning and decision-making; (*ii*) training with our limited dataset would not yield comparable performance; (*iii*) training from scratch demands substantial computing resources and time.

We plan to use LLaVA [13], a state-of-the-art open-source LMM, as our base model. We then fine-tune the model with our data which incorporates (*i*) structured data, including sensor data logs, (*ii*) unstructured data, including aerial photography and video, and (*iii*) semi-structured data such as network packets, where headers have well-defined formats but payloads are unstructured.

Given the high variance in modalities of our input data, fine-tuning the model is a complex task.

To address this complexity, we follow prior work [9] and adopt a joint pre-training model that correlates and exploits shared semantics between modalities. In our scenario, a universal correlation across all modalities is the use of timestamps, while different subsets of modalities may share other specific correlations. This allows us to fuse data collected from different sources and create a cohesive training set.

2.3 Inference

Model Compression. State-of-the-art LMMs, with hundreds of billions of parameters, require substantial memory and powerful CPU/GPU resources for inference, making them impractical for edge or some fog UAVs. Given the severe resource constraints on edge UAVs, we plan to adopt a combination of state-of-the-art approaches to effectively compress the model [25], including task-specific model pruning, knowledge distillation, and quantization.

Moreover, we will optimize both the input and output. Specifically, for the input, we apply quantization and remove redundant or unnecessary information; for the output, we will standardize its format and build a pipeline to transform it into specific, executable low-level actions to control the UAV and mitigate intrusions at all levels.

Scheduling Model Inference. Given the hierarchical UAV swarm architecture and varying performance capabilities of LMMs at each level, edge and fog UAVs have multiple inference options: local inference or sending data to the upper layer for more powerful LMM inference. The upper layer can make more coordinated decisions using data from multiple sources, such as other edge UAVs. A hybrid approach can be used: quickly responding to security events based on local inference while awaiting upper-layer decisions to verify or enhance mitigation. This trade-off involves inference latency, quality, and resource constraints.

The trade-off involves: (i) communication latency, which varies with UAV positions; (ii) differing inference times across tiers due to varying hardware and model sizes, with higher tiers having more powerful processing units and larger models; (iii) power consumption, crucial for edge UAVs with limited battery life, requiring efficient management of energy for data transmission and local inference; and (iv) the accuracy of upper-tier LMMs, which are larger and better at identifying security issues and proposing mitigations. Balancing these factors is crucial for securing UAV swarm operations while maintaining efficiency. We will evaluate trade-offs through comprehensive experiments and mathematical modeling. Recognizing the dynamic nature of real-world environments, such as varying communication in complex terrain and non-linear performance gaps in LMMs, we propose implementing a Mixture of Experts (MoE) model [11] to capture these dynamics and optimize inference decisions efficiently.

2.4 Evaluation Plan

Test Facility. University of Michigan hosts the M-Air test facility [3], a 10K sqft, four-story building designed for testing UAVs in diverse weather conditions, from −30C to 38C, including rain, sleet, wind, and snow. This testing facility is integrated with the university's data infrastructure, using a set of high-resolution motion capture cameras for experimentation. This provides an ideal environment for deploying SSRC's hardware and software. We will replicate UAV scenarios using Saluki v3.0 devices and conduct periodic M-Air tests after in-lab simulations. Data collection will include sensory inputs (e.g., camera, IMU), UAV communication, and operational logs (e.g., syscall traces), curated under various weather conditions for real-world LMM training. Additionally, we plan to use existing TII datasets to bootstrap our training.

Model Training. We will evaluate the joint pre-training model from various perspectives, including accuracy, robustness, and generalization ability across different datasets and tasks. Additionally, we will examine the effectiveness of different data sources under various scenarios by utilizing tagged data in a supervised learning framework, providing insights into both real-world and potential synthetic attack scenarios.

Model Compression. We will assess performance in terms of intrusion detection accuracy and mitigation effectiveness. We will thoroughly explore the trade-off space between compressed model size and quality by comparing the compressed model's performance against the full model to determine the impact on accuracy and efficiency.

Scheduling Model Inference. We will simulate dynamic scenarios to evaluate model inference capabilities, including varying UAV positions, environmental conditions, and UAV types. The focus will be on the effectiveness and accuracy of our MoE model's inference options. The primary evaluation metric is defense effectiveness. For each scenario, we will determine the optimal inference option by iterating over all possibilities and use this as the ground truth for comparison.

Real-World Scenario Evaluation. We will create real-world threat scenarios, such as software exploits and hand-crafted malware samples, and demonstrate how well the intrusion detection system can detect and respond in real time. Our end goal is to deliver a system and scenario at TRL4, with real hardware-in-the-loop experimentation with SSRC devices, as well as a live demo that showcases zero-trust mission.

2.5 Discussion

LMMs and existing anomaly detection systems. There are four potential design options: (i) replace all existing anomaly detection systems with LMMs; (ii) retain some existing systems while using LMMs for complementary tasks; (iii) integrate LMMs to monitor and potentially override the decisions of existing systems when they have high confidence; (iv) use the output of existing systems as input for LMMs to reference. In our exploration, we will initially focus on the first approach but will evaluate all four options to determine the most effective solution.

Limitations. The energy consumption of LMMs during inference in UAVs poses a potential limitation, especially given the constrained battery capacities of edge/fog UAVs. For example, inference with LLMs (e.g., Bloom, GPT-3) can consume up to 3.96 Wh per request [5]. For a UAV continuously processing sensory inputs and communication logs, this can rapidly drain its battery. These concerns can be partially or fully addressed by employing model compression (2.3), e.g., quantization, parameter pruning, tasks offloading, and others.

Another limitation of LMM-based intrusion detection is its vulnerability to adversarial machine learning attacks. Recent studies [7,15,17,18,23,26] show that adversaries can bypass LMMs using both visual [7,17] and textual [15,18] inputs. Attackers could manipulate sensory inputs or system logs to deceive the LMM, leading to false negatives or even system manipulation. For example, injecting carefully crafted perturbations into a camera feed could cause the LLM to misinterpret the environment or miss a real threat [8]. Similarly, manipulating system logs could trick the LMM into classifying malicious activities as benign. Given the reliance on real-time data analysis for intrusion detection, these attacks could severely compromise the security of the UAV swarm. Addressing these issues will require exploring robust defense mechanisms, such as adversarial training or anomaly detection techniques tailored to LMMs.

References

1. Downed, U.S.: Drone points to cyber vulnerabilities. https://www.washingtonpost.com/politics/2023/03/16/downed-us-drone-points-cyber-vulnerabilities/
2. Log analytics powered by AI. https://github.com/logpai
3. M-Air — michigan robotics. https://robotics.umich.edu/about/mair/
4. Al-Haija, Q.A., Badawi, A.A.: High-performance intrusion detection system for networked uavs via deep learning. Neural Comput. Appl. **34**(13), 10,885–10,900 (2022).URL https://doi.org/10.1007/S00521-022-07015-9
5. Argerich, M.F., Patiño-Martínez, M.: Measuring and improving the energy efficiency of large language models inference. IEEE Access **12**, 80,194–80,207 (2024). https://doi.org/10.1109/ACCESS.2024.3409745
6. Astaburuaga, I., Lombardi, A., La Torre, B., Hughes, C., Sengupta, S.: Vulnerability analysis of ar. drone 2.0. In: CCWC'19
7. Carlini, N., et al.: Are aligned neural networks adversarially aligned? (2024).URL https://arxiv.org/abs/2306.15447
8. Cui, X., Aparcedo, A., Jang, Y.K., Lim, S.N.: On the robustness of large multimodal models against image adversarial attacks. In: Proceedings of the IEEE/CVF Conference on Computer Vision and Pattern Recognition (CVPR),24, pp. 625–24,634 (2024)
9. D'Alessandro, M., Calabrés, E., Elkano, M.: A modular end-to-end multimodal learning method for structured and unstructured data. CoRR **abs/2403.04866** (2024).URL https://doi.org/10.48550/arXiv.2403.04866

10. Dey, V., Pudi, V., Chattopadhyay, A., Elovici, Y.: Security vulnerabilities of unmanned aerial vehicles and countermeasures: An experimental study. In: VLSID'18, pp. 398–403. IEEE (2018)
11. Jacobs, R.A., Jordan, M.I., Nowlan, S.J., Hinton, G.E.: Adaptive mixtures of local experts. Neural Comput. **3**(1), 79–87 (1991). URL https://doi.org/10.1162/neco.1991.3.1.79
12. Liu, F., Wen, Y., Zhang, D., Jiang, X., Xing, X., Meng, D.: Log2vec: A heterogeneous graph embedding based approach for detecting cyber threats within enterprise. In: CCS'19, pp. 1777–1794 (2019)
13. Liu, H., Li, C., Wu, Q., Lee, Y.J.: Visual instruction tuning. Adv. Neural Inf. Proc. Syst. **36** (2024)
14. Liu, X., Wang, Z., Wu, Y., Miao, Q.: Segcn: a semantic-aware graph convolutional network for UAV geo-localization. IEEE J. Sel. Top. Appl. Earth Obs. Remote. Sens. **17**, 6055–6066 (2024). URL https://doi.org/10.1109/JSTARS.2024.3370612
15. Liu, Y., et al.: Jailbreaking chatgpt via prompt engineering: an empirical study (2024).URL https://arxiv.org/abs/2305.13860
16. Mohsan, S.A.H., Khan, M.A., Noor, F., Ullah, I., Alsharif, M.H.: Towards the unmanned aerial vehicles (UAVS): a comprehensive review. Drones **6**(147) (2022). https://doi.org/10.3390/drones6060147
17. Qi, X., Huang, K., Panda, A., Henderson, P., Wang, M., Mittal, P.: Visual adversarial examples jailbreak aligned large language models (2023).URL https://arxiv.org/abs/2306.13213
18. Rao, A., Vashistha, S., Naik, A., Aditya, S., Choudhury, M.: Tricking LLMS into disobedience: formalizing, analyzing, and detecting jailbreaks (2024).URL https://arxiv.org/abs/2305.14965
19. Rugo, A., Ardagna, C.A., Ioini, N.E.: A security review in the uavnet era: Threats, countermeasures, and gap analysis. ACM Comput. Surv. **55**(2), 21:1–21:35 (2023). URL https://doi.org/10.1145/3485272
20. Uludag, M.K., Veksler, M., Yilmaz, Y., Akkaya, K.: Deceptive skies: Leveraging gans for drone sensor data falsification. In: SAC'24 (2024)
21. Wang, D., Li, S., Xiao, G., Liu, Y., Sui, Y.: An exploratory study of autopilot software bugs in unmanned aerial vehicles. In: ESEC/FSE'21
22. Wang, D., et al.: An exploratory investigation of log anomalies in unmanned aerial vehicles. In: Proceedings of ICSE'24, pp. 1–13 (2024)
23. Wei, A., Haghtalab, N., Steinhardt, J.: Jailbroken: how does LLM safety training fail? (2023). URL https://arxiv.org/abs/2307.02483
24. Wei, X., Sun, C., Li, X., Ma, J.: GNSS spoofing detection for UAVS using doppler frequency and carrier-to-noise density ratio. J. Syst. Archit. (2024)
25. Zafrir, O., Larey, A., Boudoukh, G., Shen, H., Wasserblat, M.: Prune once for all: sparse pre-trained language models. CoRR **abs/2111.05754** (2021).URL https://arxiv.org/abs/2111.05754
26. Zou, A., Wang, Z., Carlini, N., Nasr, M., Kolter, J.Z., Fredrikson, M.: Universal and transferable adversarial attacks on aligned language models (2023). URL https://arxiv.org/abs/2307.15043

Open Access This chapter is licensed under the terms of the Creative Commons Attribution 4.0 International License (http://creativecommons.org/licenses/by/4.0/), which permits use, sharing, adaptation, distribution and reproduction in any medium or format, as long as you give appropriate credit to the original author(s) and the source, provide a link to the Creative Commons license and indicate if changes were made.

The images or other third party material in this chapter are included in the chapter's Creative Commons license, unless indicated otherwise in a credit line to the material. If material is not included in the chapter's Creative Commons license and your intended use is not permitted by statutory regulation or exceeds the permitted use, you will need to obtain permission directly from the copyright holder.

SHIELD: Swarm-Enabled Hierarchical Intelligent Edge Defense for Drone Swarms

Tamoghna Sarkar and Bhaskar Krishnamachari

Department of Electrical and Computer Engineering, University of Southern California, Los Angeles, CA 90007, USA
{tsarkar,bkrishna}@usc.edu

Abstract. As autonomous drone swarms are increasingly deployed in high-stake environments, they become prime targets for advanced cyber threats. Traditional centralized security mechanisms fail to meet the demands of decentralized systems, which require low latency, high resilience and adaptability. SHIELD (Swarm-Enabled Hierarchical Intelligent Edge Defense) introduces a multi-layered defense system tailored to these needs, integrating Edge AI, Federated Learning (FL), Generative AI (GenAI), and Intra-Cluster Peer-to-Peer (P2P) Communication.

At the edge, drones autonomously detect and neutralize threats in real-time even when communication with higher layers is compromised. The fog layer coordinates swarm intelligence by securely aggregating updates and distributing models, while the cloud layer uses FL and GenAI to continuously refine the global threat detection models based on real and synthetic scenarios. In cases of prolonged disconnection from the cloud, SHIELD transitions to intra-cluster P2P collaboration, enabling localized responses and coordination among drones.

This architecture empowers drones to dynamically adapt to emerging threats, providing a resilient and scalable defense system that maintains operational continuity in high-risk environments. By redefining security for autonomous systems, SHIELD enhances the reliability of drone operations in sectors like critical infrastructure and disaster response.

Keywords: Cybersecurity · FL · GenAI · Autonomous systems · Swarm intelligence

1 Related Work

Drone swarms are increasingly utilized in various fields due to their ability to perform complex tasks efficiently and autonomously[1]. In critical infrastructure surveillance[2], drone swarms monitor and protect systems such as power grids and pipelines, providing real-time data for maintenance, though attacks like Distributed Denial-of-Service (DDoS) [3] can compromise their effectiveness. For disaster response and emergency services [4], drone swarms assist in search and rescue, damage assessment, and supply delivery in inaccessible areas, but communication link loss [5] can hinder relief efforts. In environmental monitoring, they enhance precision by collecting data on environmental changes [6]. In commercial, industrial applications [7], attacks can cause operational disruptions and safety risks. Yahya et al.[8] explore FL in UAV-enhanced centralized networks, focusing on joint coverage and convergence time optimization. Fu et al. [9] illustrate their UAV-based FL system for expanding network coverage, emphasizing resource allocation efficiency but not addressing resilience during cloud disconnection or critical security concerns. Yao et al. [10] investigate energy-efficient FL in Internet of Drones (IoD) networks, focusing on reducing energy consumption for FL tasks. Cui et al. [11] present consensus control for UAV systems under Byzantine attacks, enhancing resilience in adversarial conditions. Towards addressing key limitations of centralized FL models, such as single points of failure, privacy issues, and node unreliability Qu et al. [12] propose a decentralized federated learning architecture (DFL-UN) for UAV networks. The architecture enables collaborative learning among UAVs without the need for a central entity, which improves robustness and efficiency.

However, none of these work incorporate layered defense architecture or integrate FL and GenAI for proactive threat mitigation.

2 Threat Landscape

Drone swarms, while powerful in their collaborative capabilities, are highly susceptible to a range of threats reducing effectiveness of the swarm, thus leading to mission failure. Below are the key threats that drone swarms face:

- **DDoS Attacks:** Disrupting coordination by overwhelming communication channels.
- **Spoofing and Hijacking:** Misleading drones by falsifying GPS signals or hijacking communication links, causing deviation from mission paths or loss of control.
- **Eavesdropping:** Intercepting or modifying sensitive data like video feeds and telemetry, compromising operations.
- **Unknown and Evolving Threats:** Introducing new attack vectors, such as advanced malware or AI-driven threats, that adapt in real-time to defenses.

Motivation: The prior work highlights the growing use of autonomous drone swarms in critical sectors, emphasizing the need for a strong, adaptive security framework to ensure safe and continuous operations against these potential threats.

3 SHIELD: The Proposed Solution

To address the variety of threats identified earlier, we propose the **SHIELD** system, a multi-layered security framework designed to protect drone swarms from sophisticated cyberattacks ensuring robust, proactive, and reactive defense during both full cloud connectivity and intermittent or prolonged disconnection.

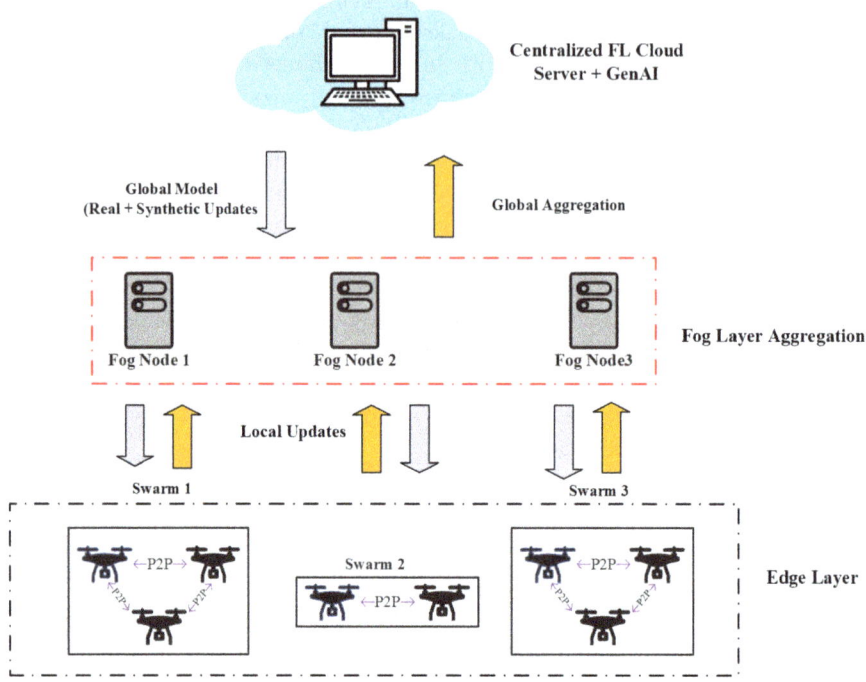

Fig. 1. Overview of the proposed SHIELD framework

SHIELD is tailored to handle current and evolving threats through a combination of *Resilient Edge AI*, *FL*, and *GenAI*. Below, we detail how SHIELD addresses these threats, while also presenting the key innovations and technical challenges.

3.1 Resilient Edge AI for Autonomous Threat Detection

SHIELD equips each drone with decentralized AI models, enabling them to detect and respond to threats in real-time without relying on a central server. By processing data locally, SHIELD minimizes the need for constant communication with the cloud, making the swarm more resilient to threats targeting network infrastructure, such as DDoS attacks.

How It Solves Threats:

- *DDoS Attacks:*
    - Detection: Localized AI processing ensures that drones remain operational even if communication channels are overwhelmed by malicious traffic.Mitigation: SHIELD implements traffic rate limiting and local prioritization of critical communications. The fog layer performs load balancing across drones to distribute resources, ensuring that vital communication is prioritized while managing the network strain during an ongoing attack.

- Mitigation: SHIELD implements traffic rate limiting and local prioritization of critical communications. The edge layer drones in SHIELD redirect unnecessary traffic into a "black hole," isolating malicious traffic from affecting the swarm.
- *Spoofing & Hijacking:*
  - Detection: Drones autonomously detect anomalies in GPS signals or communication patterns and respond immediately to mitigate hijacking or spoofing attempts.
  - Mitigation: Upon detecting spoofing or hijacking attempts, SHIELD initiates a cross-verification of sensor data, using multiple sensor modalities to validate the location and integrity of drones. If inconsistencies are detected, drones revert to backup navigation algorithms or predefined safe behavior modes to prevent mission disruption.

Modeling Approach: The decentralized AI models deployed at the edge will leverage lightweight neural networks, optimized for real-time threat detection on constrained drones and analyze network data, sensor inputs, and communication patterns to identify and mitigate threats. This approach draws on our previous work [13].

Key Innovation: The deployment of decentralized AI directly on the drones enables real-time threat detection at the edge, reducing latency and the dependency on external infrastructure.

Technical Challenge: Developing lightweight models that can run on drones while detecting and responding to complex threats in real-time.

3.2 Fog Layer and P2P Intra-Cluster Collaboration for Resilient Operations

In SHIELD's architecture, the fog layer and intra-cluster P2P communication ensure resilience and continuity, whether cloud connectivity is available or lost.

Fog Layer as a Secure Intermediary and Aggregator: The fog layer acts as a secure intermediary between drones and the cloud:

- *With Cloud Connectivity:* The fog layer aggregates model updates from drones and relays them to the cloud, reducing communication overhead and easing the burden on drones. It acts as a secure communication relay, ensuring efficient data transfer to the cloud for global model refinement.
- *Without Cloud Connectivity:* If the cloud connection is lost, the fog layer distributes a cached global model, maintaining some level of swarm coordination. After that, the fog layer's role diminishes, and it no longer engages in model training or decision-making.

P2P Intra-Cluster Collaboration: If cloud connectivity remains unavailable after distributing the cached model, drones rely on intra-cluster P2P communication. This allows drones within the same cluster to share real-time observations, threats, and coordinate local actions such as rerouting and mitigating risks, even without fog or cloud support.

How It Solves Threats and Key Innovation:

– *During Cloud Disconnection:* The fog layer distributes the cached global model, ensuring temporary swarm coordination.
– *Local Threat Mitigation:* Intra-cluster P2P collaboration enables continued operations in localized environments by sharing threat intelligence and coordinating responses during cloud disconnection.

Technical Challenges:

– *Fog Layer:* Managing communication and model distribution efficiently while transitioning between normal operations and cloud disconnection phases.
– *P2P Communication:* Ensuring reliable and secure communication within clusters without central control.

3.3 FL for Swarm-Level and Global Intelligence

SHIELD leverages FL to enable decentralized intelligence sharing. Model updates from drones are aggregated at the fog layer and then relayed to the cloud, where the global model is refined. This approach allows SHIELD to harness collective swarm intelligence without sharing raw data, thereby preserving security and privacy.

How It Solves Threats:

– *Eavesdropping & Data Tampering:*
  - Detection: FL reduces the transmission of sensitive data by sending only model updates, minimizing the risk of intercepted communications.
  - Mitigation: SHIELD employs encryption at the edge and fog layers to secure all transmitted data. If eavesdropping is suspected, drones switch to a more secure, low-bandwidth communication mode to continue safe operation with minimal exposure to sensitive information.
– *Evolving Threats:*
  - Detection: The FL process ensures that global models are continuously updated based on swarm-level intelligence, keeping the swarm resilient against emerging threats.
  - Mitigation: SHIELD anticipates potential threats using predictive models generated by GenAI, and proactively hardens drones by updating defense mechanisms before attacks occur. This allows SHIELD to handle new attack vectors by preemptively adapting threat responses in real time.

Key Innovation: Hierarchical FL, with intelligence aggregation at both the fog and cloud levels, allows SHIELD to operate securely and efficiently.

Technical Challenge: Aggregating model updates accurately and efficiently in a decentralized environment while maintaining low latency and ensuring consistency across multiple drone swarms.

3.4 GenAI for Continuous Learning and Adaptation

GenAI in the cloud generates synthetic attack scenarios that simulate new and evolving threats. These scenarios are used to continuously train the global AI model, ensuring that SHIELD stays ahead of adversaries who may employ novel futuristic attack vectors.

How It Solves Threats and Key Innovation:

- *Unknown & Evolving Threats:* By continuously generating synthetic attack scenarios, GenAI ensures that SHIELD's AI models are prepared to handle new, unseen threats that may arise in the future.
- *Proactive Defense:* GenAI-driven training allows SHIELD to anticipate and counteract attacks before they occur in the real world.

Technical Challenge: Balancing the complexity of synthetic attack scenarios with the computational and bandwidth constraints of the system to ensure that the generated data is both effective and manageable (Figs. 1 and 2).

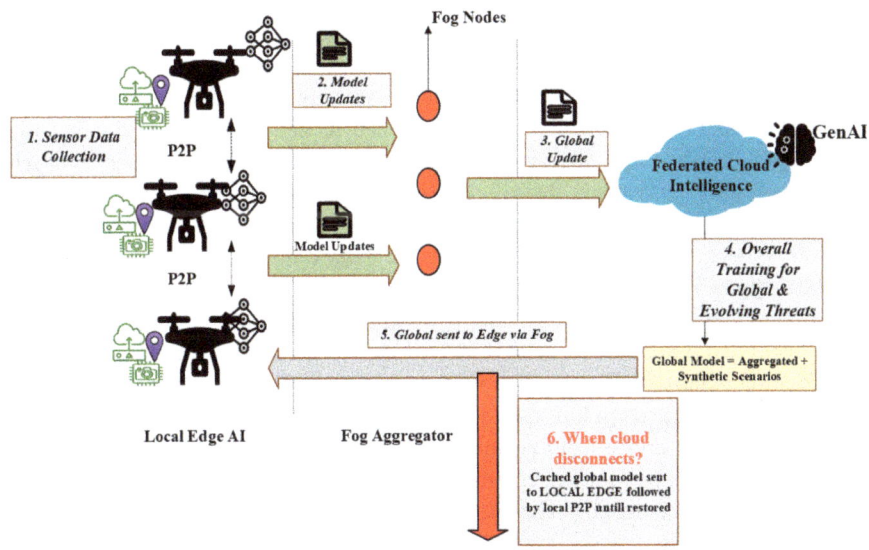

Fig. 2. Overview of SHIELD Workflow.

4 Evaluation

SHIELD's training data will be generated through a combination of real-world data collected by drones and synthetic data produced by GenAI.

4.1 Real-World Data

Drones will continuously collect data from onboard sensors, including cameras, GPS, IMUs, and network traffic monitors, to capture both normal operations and potential threats such as DDoS attacks, GPS spoofing, and environmental anomalies.

4.2 Synthetic Data

The cloud will generate synthetic data using *Generative Adversarial Networks (GANs)*, *Variational Autoencoders (VAEs)*, and adversarial examples. This data will simulate complex cyberattacks, environmental challenges, and evolving adversarial behaviors that may not be frequently encountered in real-world operations.

4.3 Data Integration

The cloud will amalgamate real-world and synthetic data into balanced datasets for model training. Initially, real-world data, collected by the drones during their operations, will be prioritized to ensure that the models are grounded in actual operating conditions and known threats. Synthetic data, generated by GenAI, will be gradually introduced to complement this real-world data, particularly to simulate novel or rare threat scenarios not frequently encountered in practice.

The balance between real-world and synthetic data will dynamically shift as the models mature and the system gains more operational experience. Early training stages will rely more heavily on real-world data, but as the models stabilize, the influence of synthetic data may increase, particularly for preparing the system for emerging or unforeseen attack vectors. The precise balance between these data types will be fine-tuned through experimentation.

To prevent overfitting and ensure robustness, the system will employ continuous feedback loops from real-world drone operations. Regular validation and retraining phases will allow the system to recalibrate the weighting of real-world versus synthetic data based on the evolving threat landscape. Additionally, adversarial training and ensemble learning techniques will be used to combine insights from both data sources, ensuring that the global models remain adaptive to both current and emerging threats.

4.4 Model Development

Edge (Drones): Lightweight *CNNs* and *RNNs*, will focus on real-time threat detection and mitigation. On-device learning will allow drones to adapt to new threats as they emerge.

Cloud: The cloud will train complex global models, such as *DNNs* and *GANs*, integrating intelligence from multiple drones to refine threat detection. Periodic updates will be pushed to the edge, ensuring drones have the latest defense mechanisms.

4.5 Hardware Platforms and Deployment

SHIELD is designed to be deployable across a range of hardware platforms typically used in drone swarms. At the edge, SHIELD's decentralized AI models will be deployed on drones equipped with lightweight, resource-constrained hardware. Examples include platforms like the NVIDIA Jetson Nano or the Raspberry Pi 4, which offer sufficient computational power for real-time threat detection while maintaining energy efficiency, critical for drone operations. These platforms support AI inference tasks through frameworks such as TensorFlow Lite or PyTorch Mobile, enabling SHIELD's real-time autonomous threat detection.

For fog nodes, more powerful hardware such as NVIDIA Jetson Xavier or Intel NUC platforms will be used. These platforms provide higher computational capabilities to aggregate model updates and facilitate secure communication between drones and the cloud.

The cloud layer (where FL and GenAI operations occur), will leverage high-performance GPU servers, such as NVIDIA A100 or V100, to efficiently handle global model training and synthetic data generation.

This hardware stack ensures scalability, as SHIELD can be deployed on a wide range of drone systems, from small, low-power drones to more robust, powerful UAVs, while maintaining seamless communication and coordination across the edge, fog, and cloud layers.

4.6 Handling Heterogeneous Data

SHIELD will employ *multi-modal learning* and data normalization techniques to handle the diverse types of data collected from various sensors such as images, time series data, GPS logs, network logs.

Simulation and Real-World Testing: SHIELD will be evaluated through simulations (e.g., NS-3, OMNeT++) to replicate various threats and real-world deployments where drones will face live cyber threats while performing tasks like surveillance and disaster response, validating system performance.

Hybrid Evaluation: Hybrid evaluations will integrate simulations with real-world testing. Simulations will generate attack patterns, tested in live scenarios to assess SHIELD's response. This iterative process ensures continuous optimization, maintaining accuracy and resilience.

Acknowledgements. This document has been edited with the assistance of ChatGPT. We certify that ChatGPT was not utilized to produce any technical content and we accept full responsibility for the contents of the paper.

References

1. Khan, M.A., Kumar, N., Mohsan, S.A.H., et al.: Swarm of UAVs for network management in 6G: a technical review. IEEE Trans. Netw. Serv. Manage. **20**(1), 741–761 (2023). https://doi.org/10.1109/TNSM.2022.3213370
2. Chen, B.-W., Rho, S.: Autonomous tactical deployment of the UAV array using self-organizing swarm intelligence. IEEE Consum. Electron. Mag. **9**(2), 52–56 (2020). https://doi.org/10.1109/MCE.2019.2954051
3. Cao, Y., Han, L., Zhao, X., Pan, X.: AccFlow: defending against the low-rate TCP DoS attack in wireless sensor networks. CoRR, vol. abs/1903.06394 (2019). http://arxiv.org/abs/1903.06394
4. Erdelj, M., Natalizio, E., Chowdhury, K.R., Akyildiz, I.F.: Help from the sky: leveraging UAVs for disaster management. IEEE Pervasive Comput. **16**(1), 24–32 (2017). https://doi.org/10.1109/MPRV.2017.11
5. Hassija, V., Chamola, V., Agrawal, A., et al.: Fast, reliable, and secure drone communication: a comprehensive survey. IEEE Commun. Surv. Tutorial **23**(4), 2802–2832 (2021). https://doi.org/10.1109/COMST.2021.3097916
6. Manfreda, S., McCabe, M.F., Miller, P.E., et al.: On the use of unmanned aerial systems for environmental monitoring. Remote Sensing **10**(4), 641 (2018). https://doi.org/10.3390/rs10040641
7. Miao, Y., Hwang, K., Wu, D., Hao, Y., Chen, M.: Drone swarm path planning for mobile edge computing in industrial Internet of Things. IEEE Trans. Industr. Inf. **19**(5), 6836–6848 (2023). https://doi.org/10.1109/TII.2022.3196392
8. Yahya, M., Maghsudi, S., Stanczak, S.: Federated Learning in UAV-Enhanced Networks: Joint Coverage and Convergence Time Optimization. arXiv preprint arXiv:2308.16889 (2023). https://arxiv.org/abs/2308.16889
9. Fu, M., Shi, Y., Zhou, Y.: Federated learning via unmanned aerial vehicle. IEEE Trans. Wireless Commun. **23**(4), 2884–2900 (2024). https://doi.org/10.1109/TWC.2023.3303492
10. Yao, J., Sun, X.: Energy-Efficient federated learning in internet of drones networks. In: IEEE 24th International Conference on High Performance Switching and Routing (HPSR), pp. 185–190 (2023). https://doi.org/10.1109/HPSR57248.2023.10147956
11. Cui, Y., Liang, Y., Luo, Q., et al.: Resilient consensus control of heterogeneous Multi-UAV systems with leader of unknown input against Byzantine attacks. IEEE Trans. Autom. Sci. Eng. (2024). https://doi.org/10.1109/TASE.2024.3420697
12. Qu, Y., Dai, H., Zhuang, Y., et al.: Decentralized federated learning for UAV networks: architecture, challenges, and opportunities. IEEE Netw. **35**(6), 156–162 (2021). https://doi.org/10.1109/MNET.001.2100253
13. Hekmati, A., et al.: Correlation-Aware neural networks for DDoS attack detection in IoT systems. IEEE/ACM Trans. Netw. 1–16 (2024). https://doi.org/10.1109/TNET.2024.3408675

Open Access This chapter is licensed under the terms of the Creative Commons Attribution 4.0 International License (http://creativecommons.org/licenses/by/4.0/), which permits use, sharing, adaptation, distribution and reproduction in any medium or format, as long as you give appropriate credit to the original author(s) and the source, provide a link to the Creative Commons license and indicate if changes were made.

The images or other third party material in this chapter are included in the chapter's Creative Commons license, unless indicated otherwise in a credit line to the material. If material is not included in the chapter's Creative Commons license and your intended use is not permitted by statutory regulation or exceeds the permitted use, you will need to obtain permission directly from the copyright holder.

Swarming Tight Interactions for Achieving Resistibility of Large Robotic Systems in Real-World Conditions

Jiri Horyna(✉) and Martin Saska

Multi-Robot Systems Group Faculty of Electrical Engineering, Czech Technical University in Prague, Technicka 2, Prague, Czech Republic
horynjir@fel.cvut.cz
https://mrs.fel.cvut.cz/

Abstract. This paper presents an autonomous swarm system designed to be an enabling technology for achieving resilience to both partial and complete dropouts of localization of individual vehicles in large teams. The challenge of creating a resilient swarm system across diverse mission types is closely tied to maintaining accurate state awareness, regardless of changing environmental conditions and external threats like jamming and spoofing of primary localization data. Leveraging purely relative measurements and onboard sensor data to ensure accurate state awareness despite intermittent localization failures is extremely important for enhancing security, resilience, and safety of cooperating systems including edge autonomous devices. The first part of this paper focuses a system designed to resist partial localization dropouts of individual robots or even subgroup of robots due to spatial unavailability of localization modalities. This system integrates robust mutual perception mechanisms and shared measurements into the closed-loop primary state estimation pipeline. Such an approach enables the swarm to continue its mission even when localization dropouts occur among a subset of edge drone agents. The second part of the paper examines state-of-the-art techniques aimed at achieving resilience in the event of a global localization dropout, relying exclusively on relative onboard measurements. This method allows the swarm to actively maintain its local constellation, enabling it to continue the mission even after the threat subsides, though at the cost of temporary formation drift. By combining these two approaches, the paper bridges the gap in enhancing the resilience of drone swarm operations, allowing them to adapt dynamically across a wide range of mission types. In our workshop presentation, we will introduce and discuss the description and results of these state-of-the-art distributed state estimation techniques, which significantly strengthen swarm system security against vulnerabilities posed by emerging threats.

Keywords: multi-robot state estimation · cooperative localization · UAV swarms

1 Introduction

Achieving secure state estimation in aerial autonomous systems across a variety of mission types and real-world environmental conditions is a challenging task. It involves designing a system that is both versatile and adaptive, as well as responsive to external threats. External localization architectures, such as the Global Navigation Satellite System (GNSS), may be unavailable or unsuitable due to issues like low accuracy and unreliability in urban environments. Moreover, without proactive security algorithms, these techniques are vulnerable to interference, jamming, or spoofing. In Unmanned Aerial Vehicle (UAV) research, various self-localization approaches, such as Simultaneous Localization and Mapping (SLAM) [1] algorithms and Visual-Inertial Odometry (VIO) [2], have been developed to stabilize and navigate robots in GNSS-denied environments. However, regardless of the primary localization source, onboard state estimation accuracy can temporarily decrease due to factors like computational singularities, environmental characteristics, or unexpected sensor malfunctions. For instance, the accuracy of methods relying on optical cameras can be significantly compromised by homogeneity in the camera image (Fig. 1c).

Our solution to these bottlenecks that prevent direct use of these techniques in zero trust multi-robot architectures is based on decentralized and distributed state estimation frameworks that combine the primary localization source with onboard observations of surrounding agents. This approach ensures stable flight even when confidence in the primary localization decreases, such as when an external threat is detected. These independent frameworks enhance the swarm system's resilience against undesirable and uncontrollable negative effects, whether they impact only a subset of agents or the entire swarm.

2 Multi-robot State Estimation

Multi-robot state estimation (MRSE) during partial degradation of primary localization performance relies on fusing primary localization data (e.g., GNSS, VIO) with Inertial Measurement Unit (IMU) readings and a robust onboard mutual perception system that estimates the distance and bearing of surrounding agents. Building on previous studies [3,4], our work presents a comprehensive distributed state estimator architecture integrated into the closed-loop state estimation and control pipelines. This architecture is grounded in modeling the movements of surrounding fog agents. The model-based estimated positions of these agents serve as floating localization anchors for the focal UAV when it faces emerging threats or singularity in onboard state estimate. The focal UAV uses these floating anchors and an onboard mutual visual perception system to estimate its state. The MRSE is further enhanced by fusing it with IMU data, ensuring robustness during agile maneuvers. Final resilience of the UAV state estimate is achieved by adaptively fusing the MRSE with the primary state estimator, based on the monitored confidence in the primary state estimate. By integrating and refining collective measurements, the system compensates

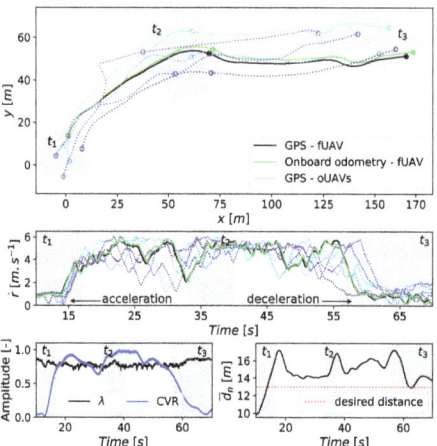
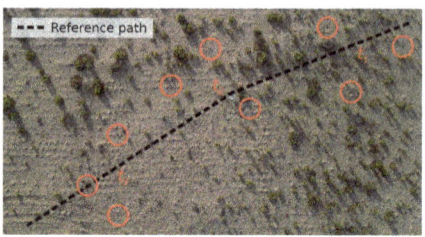

(a) Swarm approaching a static goal. Swarm UAVs' positions, velocities, and qualitative parameters are displayed. $\bar{d}_n = 15.12\,\text{m}$ with standard deviation 5.07 m. GNSS is used as a ground truth.

(b) Flocking of three UAVs with the proposed estimator. Timestamps within the flight are: $t_1 = 10\,\text{s}$, $t_2 = 55\,\text{s}$, $t_3 = 100\,\text{s}$.

(c) A swarm of six UAVs (yellow) using the proposed approach. The group velocity of $5\,\text{m s}^{-1}$ was reached, while the swarm stayed coherent without reliance on GNSS and communication.

Fig. 1. Demonstration of MRSE approach in GNSS-denied feature-poor environments. Fusion coefficients are determined onboard according to the confidence in the accuracy of the primary state estimation approach of a particular UAV.

for individual inaccuracies, thereby enhancing overall performance. To further improve the robustness of surrounding agents' model, we incorporate an approach for estimating the immeasurable velocities of surrounding UAVs based on observed swarming behavior, making communication an optional modality.

3 Swarming Without an Anchor

State estimation during global loss of primary localization performance is designed to enhance the resilience of swarm systems by proactively responding to known or emerging threats, relying solely on relative measurements. Our method, called Swarming Without an Anchor (SWA) integrates decentralized state estimation techniques with robust mutual perception mechanisms and onboard sensor data to maintain accurate state awareness despite intermittent localization failures. It employs an onboard mutual perception system to determine the relative positions of neighboring agents, which are used to define an unambiguous reference frame within the local constellation. This unambiguous definition of a floating reference frame enables precise estimation of the focal UAV's state relative to the local constellation without the dependence on any localization anchor fixed to the environment. Disturbances affecting individual fog drones are mitigated through a distributed high-level control law, ensuring stability among

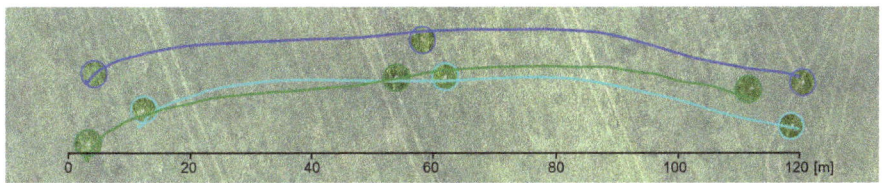

(a) Positions of UAVs with respect to the environment during the drift of the formation.

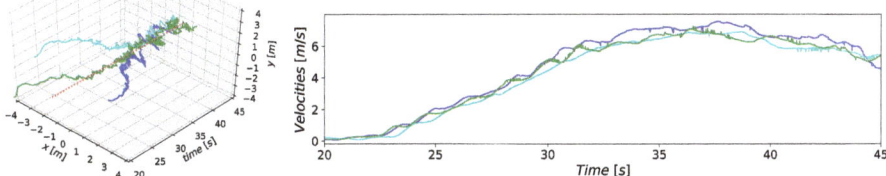

(b) Left: positions of UAVs converge to the origin of their control frames. Right: ground velocities of UAVs.

Fig. 2. Experiment with 3 UAVs during a global localization dropout. Despite the swarm drifting, the constellation is maintained through state estimation using relative measurements.

surrounding agents. Our solution demonstrates how a robot swarm adapts more effectively to localization dropouts than a single agent, which is particularly crucial in GNSS-denied environments lacking significant localization features, as shown in Fig. 2b. Compared to a single-drone scenario, the proposed purely relative state estimation and control strategy SWA allows a team of UAVs to continue their mission even during global localization dropouts or during external attempt to break through state estimation security. Although the overall multi-robot system is not jointly observable with respect to the environment, leading to formation drift, any drift other than uniform translation of the swarm as a whole is attenuated. This behavior facilitates velocity consensus among the UAVs, effectively addressing the double integrator synchronization problem.

4 Experimental Section

The experiments were designed to validate the system's functionality in realistic GNSS-denied, feature-poor environments, a critical step for identifying and addressing system bottlenecks in future development. A F4F F450 quadcopter equipped with a Pixhawk autopilot, a rangefinder for precise altitude control, and a downward-facing camera for VIO was used (for more details on the UAV setup, see [5]). The system also featured an onboard relative perception system to estimate the positions of surrounding agents relative to the focal agent. In Fig. 1, UVDAR [6,7] was employed for mutual perception; however, any onboard mutual perception system can be integrated with the introduced state estimation approaches. All control [8], estimation, and planning computations were conducted onboard an Intel NUC-i7 with embedded WiFi for optional communication, utilizing the Robot Operating System (ROS).

The introduced MRSE[1] strategy was successfully tested in feature-poor environments, such as a desert and plain grass field, under conditions of partial state estimation performance degradation caused by uniformity in the VIO camera image. Specifically, as shown in Fig. 1c and Fig. 1a, the UAVs used VIO fused with the MRSE for self-localization and UVDAR for mutual perception and collision avoidance. The swarm covered over 200 m with an average velocity of $5.0, m.s^{-1}$. The rapid movement above the grass surface made VIO feature tracking more challenging, reducing confidence in VIO localization and leading to increased reliance on the MRSE, as indicated by a lower coefficient λ.

The resilience against the entire global localization loss using the introduced SWA[2] approach was tested using three UAVs hovering at a desired height (see Fig. 2a). Initially, the UAVs relied on GNSS for self-localization. Then, all onboard state estimators were switched to the introduced state estimation method, which uses only relative measurements. This setup simulated the emerging threat of GNSS jamming, leaving the UAVs without world frame position and velocity data. Despite this, the swarm flew over 100 m (Fig. 2)a, with drift velocity increasing to $7\,m.s^{-1}$ (Fig. 2)b, and successfully completed the experiment without collisions, maintaining cohesion through relative measurements alone. The UAVs' relative positions converged to the origin of their floating frames (Fig. 2)b, in line with the intended design of the state estimator and feedback control approach.

5 Conclusion and Future Work

This paper introduces an autonomous swarm system aimed at improving resilience in large UAV teams, especially in situations where individual vehicle localization is partially or entirely compromised. By leveraging purely relative measurements and onboard sensor data, we have showcased a robust method for sustaining accurate state awareness, even in the presence of environmental challenges and external threats like jamming and spoofing.

As future work, we propose an integration of the developed system into a fully secured multi-UAV framework, with an emphasis on identifying edge cases where no sensory information is available. Additionally, we plan to incorporate AI algorithms for threat detection and mitigation, and design coordinated mechanisms that enable a collective swarm response to threats. The system will also be generalized to accommodate heterogeneous multi-robot teams with varying dynamics. Secured communication will be integrated, ensuring that data is used only when it is both available and trustworthy.

---

[1] Video of MRSE approach in desert: http://mrs.felk.cvut.cz/iros-2022-estimation of MRSE approach in grass field: https://mrs.fel.cvut.cz/ral-2023-demo.
[2] Video of SWA approach: https://youtu.be/kPiOdsPKh-U?si=HHr9xecnXYJaZM2S.

References

1. Taketomi, T., et al.: Visual SLAM algorithms: a survey from 2010 to 2016. IPSJ Trans. Comput. Vision Appl. **9**(1), 1–11 (2017)
2. Weinstein, A., et al.: Visual inertial odometry swarm: an autonomous swarm of vision-based quadrotors. IEEE RA-L **3**(3), 1801–1807 (2018)
3. Horyna, J., et al.: Decentralized multi-robot velocity estimation for UAVs enhancing onboard camera-based velocity measurements, In: IEEE/RSJ IROS (2022)
4. Horyna, J., et al.: Fast swarming of UAVs in GNSS-denied feature-poor environments without explicit communication. IEEE Robot. Autom. Lett. **9**(6), 5284–5291 (2024)
5. Hert, D., et al.: MRS drone: a modular platform for real-world deployment of aerial multi-robot systems. J. Intell. Robot. Syst. **108**, 1–34 (2023)
6. Walter, V., et al.: UVDAR system for visual relative localization with application to leader-follower formations of multirotor UAVs. IEEE RA-L **4**(3), 2637–2644 (2019)
7. Walter, V., et al.: Mutual localization of UAVs based on blinking ultraviolet markers and 3D time-position Hough transform, In: 14th IEEE CASE, 2018
8. Baca, T., et al.: The MRS UAV system: pushing the frontiers of reproducible research, real-world deployment, and education with autonomous unmanned aerial vehicles. J. Intell. Robot. Syst. **102**(26), 1–28 (2021)

Open Access This chapter is licensed under the terms of the Creative Commons Attribution 4.0 International License (http://creativecommons.org/licenses/by/4.0/), which permits use, sharing, adaptation, distribution and reproduction in any medium or format, as long as you give appropriate credit to the original author(s) and the source, provide a link to the Creative Commons license and indicate if changes were made.

The images or other third party material in this chapter are included in the chapter's Creative Commons license, unless indicated otherwise in a credit line to the material. If material is not included in the chapter's Creative Commons license and your intended use is not permitted by statutory regulation or exceeds the permitted use, you will need to obtain permission directly from the copyright holder.

RoboMesh: Swarm-Based Orchestration for Secure Multi-robot Systems

Reyhaneh Rabaninejad[✉] and Antonis Michalas

Tampere University, Tampere, Finland
{reyhaneh.rabbaninejad,antonios.michalas}@tuni.fi

Abstract. In the dynamic realm of multi-robot systems, secure decision-making and efficient resource utilization are paramount. This work introduces RoboMesh, a decentralized orchestration framework leveraging swarm intelligence and generative AI to optimize application deployment across the cloud-to-edge continuum. By implementing self-organized swarms, we enhance secure decision-making processes within multi-robot environments, enabling adaptive responses to fluctuating conditions and threats. Knowledge and trust, essential for the operation of the orchestration space, are managed through blockchain-based solutions and emerging technologies like Self-Sovereign Identities (SSI). The system's end-to-end security is ensured through advanced cryptographic techniques and privacy-preserving data analytics. Additionally, a digital twin runs parallel to the physical system, providing predictive feedback to further improve performance. RoboMesh embodies a cutting-edge approach to enhancing the resilience and efficiency of multi-robot systems in complex environments.

Keywords: Decentralized Orchestration · Swarm Intelligence · Generative AI · Resource Optimization · Multi-Robot Systems

1 Introduction

Multi-robot systems operating in resource-constrained environments require secure, efficient orchestration to manage resources dynamically. Traditional centralized models often fall short in these scenarios, presenting risks such as single points of failure and limited adaptability [4]. RoboMesh introduces a decentralized orchestration framework that leverages swarm intelligence and generative AI, facilitating seamless decision-making and resource management in real time.

Orchestration, as defined by Jiang et al. [5], involves coordinating and managing physical and computational resources to meet application requirements. While Tomarchio et al. [11] discuss this in the context of cloud environments, and Costa et al. [3] within fog computing, none of these can fulfil completely the highly dynamic and complex requirements imposed by the Cloud-to-Edge continuum.Existing orchestration solutions often rely on centralized control [12], which can lead to bottlenecks, single points of failure, and security vulnerabilities.

These approaches struggle with the dynamic, distributed nature of edge environments, where local data processing and rapid adaptation to changing conditions are essential.RoboMesh addresses these challenges through decentralized swarm intelligence, enabling seamless decision-making and resource management. The framework utilizes blockchain-based trusted solutions and emerging technologies like Self-Sovereign Identities (SSI) aka Decentralized Identifiers (DID) [1,8] to maintain knowledge and trust within the orchestration space. Decentralized storage networks are integrated to facilitate secure and scalable data sharing among swarm nodes, ensuring data availability and integrity even in dynamic environments [7,10]. Additionally, it employs state-of-the-art cryptographic techniques and privacy-preserving data analytics [2,9] to ensure end-to-end security.

2 RoboMesh Overview

RoboMesh represents a new generation of orchestration tools that dynamically adapt to the needs of distributed, multi-robot environments, ensuring efficient, secure, and resilient operations.Specifically, our framework employs decentralized swarm-based orchestration framework where individual robots act as autonomous agents. Each agent utilizes generative AI models to assess local conditions and make informed decisions, contributing to the collective behavior of the swarm. This decentralized approach enhances security and resilience by distributing decision-making processes, reducing vulnerabilities associated with centralized systems. We incorporate advanced cryptographic techniques to safeguard communications and ensure data integrity across the swarm network.

This framework addresses the decentralized nature of multi-robot systems, which consist of heterogeneous resources across multiple domains, each with varying capacities and availability. Intelligent orchestration mechanisms are essential for managing robots within this complex environment. RoboMesh provides a high-level, interoperable descriptor that encapsulates multi-robot platform topology, constraints, optimization goals, and performance needs.

By abstracting the intricacies of the cloud-to-edge continuum, RoboMesh enables the deployment of microservices across diverse infrastructures while ensuring security and trust. Building on these essential aspects, Fig. 1 presents a high-level architecture that illustrates the overall vision of RoboMesh and its functionality. Structurally, system architecture comprises four key components: (1) Multi-robot platformfocused on platform specification, (2) Orchestration Spacemanaging core orchestration functions, (3) Trusted Knowledge Managementensuring secure system-wide knowledge management, and (4) Resource Layerrepresenting resources across cloud and non-cloud environments. The technical specification of these components is detailed in the next section.

3 RoboMesh Components Specification

3.1 Multi-robot Platform

A multi-robot platform specification within the RoboMesh framework outlines the essential components and capabilities required for effective collaboration

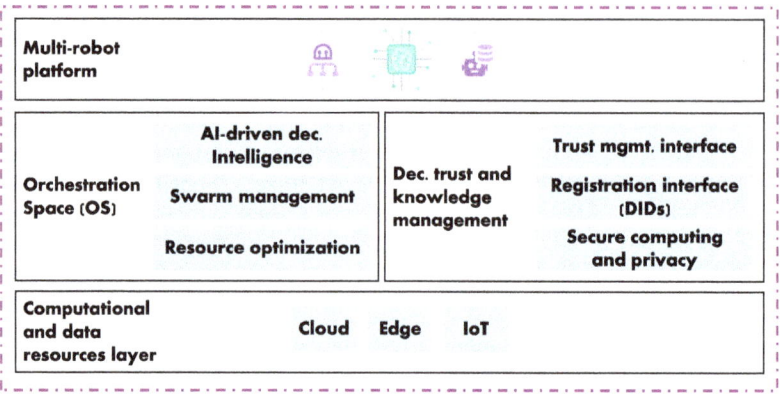

Fig. 1. High-Level Architecture of the RoboMesh Framework.

among a swarm of robots. This specification includes hardware configurations, communication protocols, and software architecture necessary to support tasks such as navigation, perception, and decentralized decision-making. The platform accommodates diverse robot types and roles, ensuring seamless interoperability and coordination. Key considerations such as processing power, sensor integration, and energy efficiency are critical for optimizing performance in dynamic, real-world scenarios. In addition to physical and operational specifications, the platform defines the algorithms and frameworks that enable distributed decision-making and real-time data processing. Advanced capabilities, including swarm intelligence and generative AI models as described in next component, allow individual robots to assess local conditions, share information, and make collective decisions that enhance the system's overall efficiency and resilience.

3.2 Orchestration Space

In RoboMesh, edge robots are registered to the Orchestration Space (OS), a distributed entity without a central access point. This space integrates decentralization, swarm intelligence, and generative AI to achieve efficient, optimized, and secure orchestration of robots within the Cloud-to-Edge ecosystem. Traditional orchestration solutions are often centralized, presenting challenges in the distributed and dynamic Cloud-to-Edge continuum where edge resources are volatile and have limited processing capabilities. RoboMesh addresses these issues by emphasizing local decision-making and collaborative interactions among multiple entities, ensuring the system's resilience and adaptability.

The core characteristic of swarm computing is the emergence of collective behavior and intelligence through interactions among distributed agents, rather than centralized control. This self-organized, scalable, and adaptable approach fits well with the dynamic nature of Cloud-to-Edge and multi-robot systems. In RoboMesh, swarms are formed based on logical proximity, determined by edge robot requirements such as CPU, memory, security, and performance needs, rather than just geographical proximity [6]. This allows for dynamic, self-organized swarms that can adjust based on required characteristics and resource availability. Orchestration Agents (OAs) within swarms handle allocation tasks and ensure efficient intra- and inter-swarm communication. Distributed AI techniques enhance swarm intelligence, optimizing swarm dynamics, coordination, and adaptation to environmental changes, ultimately ensuring secure and efficient operations across the multi-robot system.

Besides, the RoboMesh OS resource optimization includes an AI-driven allocation schema for robots' microservices, emphasizing energy efficiency across available resources. This schema considers detailed technical and economic inputs about distributed devices and robots, enabling cross-layer energy optimization. It addresses hardware, software, and networking energy issues, alongside robot-specific concerns like urgency. Due to the computational complexity and dynamic nature of the multi-robot environment, continuous learning approaches are employed to enhance the optimization process.

3.3 Decentralized Trust and Knowledge Management

Achieving trust in a decentralized environment is a complex challenge. RoboMesh registration interface (Fig. 1) addresses this by generating verifiable credentials and proofs such as proof of presence, proof of location, and proof of computing capabilities during DID issuance, which are crucial for establishing trust within the system. Utilizing Zero-Knowledge proofs across all application levels, RoboMesh prioritizes trust as a fundamental component. To support this, RoboMesh incorporates a blockchain-based decentralized knowledge and trust infrastructure, responsible for knowledge management, transparency, and trust assurance among system components, the distributed resource layer, and multi-robot platform.

The initial investigation focuses on developing formal models of trustworthiness to ensure dependable interactions among entities and services within the Cloud-to-Edge continuum. By implementing blockchain-based trust management solutions using SSI and DID, RoboMesh enables privacy-preserving identity and role management. Smart contracts and decentralized oracles manage resource and robot descriptions, system interactions, and decision-making processes, allowing both resources and robots to be discoverable based on contextual attributes and trust factors. This ensures full transparency, traceability, and secure, verifiable system.

Furthermore, the RoboMesh platform ensures comprehensive security and privacy through advanced encryption techniques such as Functional Encryption (FE) and Hybrid Homomorphic Encryption (HHE), allowing for the analysis of encrypted data in a privacy-preserving manner. It employs a blockchain-based FE mechanism for decentralized trust management, ensuring transparency among system components and resources.

3.4 Resource Layer

In the RoboMesh concept, a resource encompasses any computational entity, including virtual machines in the cloud, physical nodes across the compute continuum, or intelligent sensors with processing capabilities. Resources may also include pre-deployed software services on dedicated hardware. These resources are heterogeneous and span different administrative domains, characterized by attributes like hardware specs, operating system, location, mobility, and battery power, which help determine their suitability for specific tasks.

A resource is considered trusted once registered using DIDs in RoboMesh registration interface, producing verifiable proofs for its attributes. These proofs are validated before a resource joins or forms a Swarm, ensuring its suitability and trustworthiness for the task at hand.

4 Use Case Scenarios

RoboMesh can play a critical role in managing the intricate network of smart city infrastructure. For instance, autonomous drones and ground robots can work together to monitor and maintain public amenities, such as street lighting, waste management, navigation in urban environments, and security systems. The decentralized orchestration framework of RoboMesh ensures that these robots can operate seamlessly, independently, and securely. If a malfunction is detected in any part of the infrastructure, RoboMesh-enabled robots can autonomously allocate tasks to rectify the issue promptly, ensuring minimal disruption to urban services. The use of advanced cryptographic techniques further secures the data and communications between these autonomous units, maintaining the integrity and reliability of the smart city's operations.

As another usecase scenario, RoboMesh can transform delivery operations by orchestrating a fleet of autonomous drones and ground robots. These robots, functioning as decentralized agents, use generative AI to assess real-time conditions and optimize delivery routes. This approach enhances delivery speed, accuracy, and overall efficiency. Each robot can autonomously navigate through urban landscapes, avoiding obstacles and selecting the best paths to ensure timely deliveries. If a delivery robot encounters a blocked road, it can quickly reroute itself and update nearby robots to prevent delays.

5 Conclusion

RoboMesh presents a novel approach to orchestrating multi-robot systems, focusing on security and resource optimization in multi-robot environments. By utilizing swarm intelligence and generative AI, our framework develops secure, autonomous, and fully decentralised decision-making and efficient resource management. The framework is based on Swarm computing principles and utilises distributed AI and self-sovereign identities for multi-robot platform management. Future work will explore developing Robomesh framework using an incremental and iterative methodology and further enhancing the framework's adaptability in diverse operational contexts.

References

1. Decentralized identity foundation (2020). https://identity.foundation/
2. Bakas, A., Michalas, A., Frimpong, E., Rabaninejad, R.: Feel the quantum functioning: instantiating generic multi-input functional encryption from learning with errors. In: IFIP Annual Conference on Data and Applications Security and Privacy, pp. 279–299. Springer (2022)
3. Costa, B., Bachiega, J., Carvalho, L.R., Araujo, A.P.: Orchestration in fog computing: a comprehensive survey. ACM Comput. Surv. (CSUR) **55**(2), 1–34 (2022)
4. Hong, C.H., Varghese, B.: Resource management in fog/edge computing: a survey on architectures, infrastructure, and algorithms. ACM Comput. Surv. (CSUR) **52**(5), 1–37 (2019)
5. Jiang, Y., Huang, Z., Tsang, D.H.: Challenges and solutions in fog computing orchestration. IEEE Network **32**(3), 122–129 (2017)
6. Lera, I., Guerrero, C., Juiz, C.: Availability-aware service placement policy in fog computing based on graph partitions. IEEE Internet Things J. **6**(2), 3641–3651 (2018)
7. Rabaninejad, R., Abdolmaleki, B., Malavolta, G., Michalas, A., Nabizadeh, A.: storna: stateless transparent proofs of storage-time. In: European Symposium on Research in Computer Security, pp. 389–410. Springer (2023)
8. Rabaninejad, R., Abdolmaleki, B., Ramacher, S., Slamanig, D., Michalas, A.: Attribute-based threshold issuance anonymous counting tokens and its application to sybil-resistant self-sovereign identity. Cryptology ePrint Archive (2024)
9. Rabaninejad, R., Bakas, A., Frimpong, E., Michalas, A.: A secure bandwidth-efficient treatment for dropout-resistant time-series data aggregation. In: 2023 IEEE International Conference on Pervasive Computing and Communications Workshops and other Affiliated Events (PerCom Workshops), pp. 640–645. IEEE (2023)
10. Rabaninejad, R., Liu, B., Michalas, A.: Port: non-interactive continuous availability proof of replicated storage. In: Proceedings of the 38th ACM/SIGAPP Symposium on Applied Computing, pp. 270–279 (2023)
11. Tomarchio, O., Calcaterra, D., Modica, G.D.: Cloud resource orchestration in the multi-cloud landscape: a systematic review of existing frameworks. J. Cloud Comput. **9**(1), 49 (2020)
12. Zhong, Z., Xu, M., Rodriguez, M.A., Xu, C., Buyya, R.: Machine learning-based orchestration of containers: a taxonomy and future directions. ACM Comput. Surv. (CSUR) **54**(10s), 1–35 (2022)

Open Access This chapter is licensed under the terms of the Creative Commons Attribution 4.0 International License (http://creativecommons.org/licenses/by/4.0/), which permits use, sharing, adaptation, distribution and reproduction in any medium or format, as long as you give appropriate credit to the original author(s) and the source, provide a link to the Creative Commons license and indicate if changes were made.

The images or other third party material in this chapter are included in the chapter's Creative Commons license, unless indicated otherwise in a credit line to the material. If material is not included in the chapter's Creative Commons license and your intended use is not permitted by statutory regulation or exceeds the permitted use, you will need to obtain permission directly from the copyright holder.

Challenge 3: – Adaptive Learning for Evolving Drone Operation

Trusting Data Updates to Drone-Based Model Evolution

Marco Anisetti[1,2], Claudio A. Ardagna[1,2], Nicola Bena[1(✉)], Ernesto Damiani[1,3], Chan Yeob Yeun[3], and Sangyoung Yoon[3]

[1] Department of Computer Science, Università degli Studi di Milano, Milan, Italy
{marco.anisetti,claudio.ardagna,nicola.bena}@unimi.it
[2] Moon Cloud srl, Milan, Italy
[3] C2PS, Computer Science Department, Khalifa University, Abu Dhabi, UAE
{ernesto.damiani,chan.yeon,sangyoung.yoon}@ku.ac.ae

Abstract. AI is revolutionizing our society promising unmatched efficiency and effectiveness in numerous tasks. It is already exhibiting remarkable performance in several fields, from smartphones' cameras to smart grids, from finance to medicine, to name but a few. Given the increasing reliance of applications, services, and infrastructures on AI models, it is fundamental to protect these models from malicious adversaries. On the one hand, AI models are black boxes whose behavior is unclear and depends on training data. On the other hand, an adversary can render an AI model unusable with just a few specially crafted inputs, driving the model's predictions according to her desires. This threat is especially relevant to collaborative protocols for AI models training and inference. These protocols may involve participants whose trustworthiness is uncertain, raising concerns about insider attacks to data, parameters, and models. These attacks ultimately endanger humans, as AI models power smart services in real life (AI-based IoT). A key need emerges: *ensuring that AI models and, more generally, AI-based systems trained and operating in a low-trust environment can guarantee a given set of non-functional requirements, including cybersecurity-related ones.* Our paper targets this need, focusing on collaborative drone swarm missions in hostile environments. We propose a methodology that supports trustworthy data circulation and AI training among different, possibly untrusted, organizations involved in collaborative drone swarm missions. This methodology aims to strengthen collaborative training, possibly built on incremental and federated learning.

Keywords: Drone Swarm · Federated Learning · Incremental Learning · Trust

1 Problem Statement

From the introduction of convolutional neural networks to large language models such as Llama, Artificial Intelligence (AI) is gaining momentum (e.g., [8,10]), with an expected economic impact of approximately $13 trillion by 2030.[1]

AI-based systems are increasingly built on scenarios where coalitions of unknown and untrusted organizations collaborate in joint tasks, by deploying fleets of AI-based devices in hostile environments. In these scenarios, system operation, including AI models training and inference, is carried out collaboratively, introducing significant challenges. First, data and models must circulate across organizations in the cloud-edge continuum [3] to support system operation [15]. Artifacts circulation increases tasks quality, but poses stronger concerns about privacy and robustness. Second, non-functional requirements (e.g., fairness, robustness) may conflict. For instance, when an ensemble of AI models is used for training-time poisoning prevention [2], the more the models in the ensemble, the higher the robustness but the larger the footprint and the security risks. Finally, functional and non-functional requirements may conflict. For instance, privacy-preserving AI models trained with Federated Learning (FL) may exhibit lower accuracy.

Our case study sits at the confluence of the aforementioned challenges: *drone swarms performing collaborative missions in hostile environments*. During their missions, drones rely on AI models and collect large amounts of data, used for (re)training upon mission completion. This collaborative scenario can consider a single organization or be extended to a federation, where federated organizations can safely (and privately) share data and AI models, to further increase mission quality.

In this context, mission safety and security are primary concerns. AI models, as well as data collected during missions, may have been poisoned [2] or may be unprotected against malicious data perturbations [17]; drones may have been compromised by adversaries or rogue owners [11], and the lack of trust relationships in collaborative and federated scenarios may hamper data and model circulation.

Our proposal aims to define a comprehensive methodology for *collaborative and resilient learning for drone swarms operating in hostile environments* and is built on three pillars:

- collaborative, distributed learning accessing data from multiple sources at different levels of trustworthiness;
- secure life cycle management and assurance of AI artifacts (data, AI parameters, and AI models); and
- semi-automated deployment of robust AI models to drones.

The remainder of this abstract is organized as follows. Section 2 introduces the system model and the reference scenario target of our proposal, which is

---

[1] https://www.mckinsey.com/featured-insights/artificial-intelligence/notes-from-the-AI-frontier-modeling-the-impact-of-ai-on-the-world-economy.

explained in Sect. 3. Section 4 describes the trust annotations at the basis of artifacts circulation. Finally, Sect. 5 draws our conclusions.

2 System Model and Reference Scenario

Our proposal considers drone missions in a collaborative AI-based IoT (AIoT) scenario, where drones belonging to different organizations execute collaborative tasks (e.g., mine detection). To this aim, the organizations jointly train an (set of) AI model that is distributed and deployed on the devices part of the collaborative drones missions. Once deployed, drones start their activities and collect data on their behavior, measurements read by their sensors on the drone internals and surrounding environment, and data on AI model inference performed during the task. Once the task ends and the drones return to the organization's premises, all collected data are analyzed, assessed, and used to collaboratively (re)train the AI models across the organizations for the next mission.

The robustness and quality of the AI models throughout the entire process play a fundamental role for the accuracy of the collaboration and must thus be strengthened and guaranteed. To this aim, AI artifacts such as training data and model parameters must be safely circulated in and cross-organizations before and after every mission, preserving their privacy and the privacy of the corresponding organizations. Artifacts circulation must obey the requirements of each organization and must support high-quality, robust AI models (re)training that balance functional and non-functional requirements.

Threat Model. Our proposal is based on a threat model where the attacker aims to decrease the quality of the collaborative task outcome or prevent task completion, by compromising the operation of AI models. To this aim, the attacker attempts to interfere with the drones to poison the data they collect, which are later used to (re)train the AI model distributed to drones. The attacker may control a subset of the area where drones operate and part of the sensors of some devices, depending on the use case. The attacker exploits this control to alter the data points read and collected by the sensors, and possibly sent to the AI models deployed on the devices (e.g., poisoning), but does not have direct access to the AI models.

Goals and Assumptions. Our proposal aims to provide offline protection against poisoning attacks implementing a robust, trustworthy, and collaborative (re)training approach at the organizations' premises. The approach supports scalability, sustainability, and avoids costly real-time detection, monitoring, and on-device (re)training. Our proposal assumes that AI artifacts can circulate before and after task completion among the participating organization(s) without being attacked; it also assumes on-board AI models integrity, through a deployment on secure hardware, such as TEE (Trusted Execution Environment) or TPM (Trusted Platform Module).

2.1 Scenarios

During their missions, drones rely on AI models and collect large amounts of data that are also used for (re)training upon mission completion. The collaboration between the drones of an organization can substantially boost mission quality, with deployed AI models benefiting from the data collected by the different drones in different missions. This collaborative scenario can be extended to a federated one, where organizations in the same federation can safely (and privately) share data and AI models to increase mission quality. This scenario is challenging from different perspectives. First, maliciously created data points can be generated by multiple sources, including hacked drones, rogue organizations in the federation, or be injected at any step of the collaborative protocols. Second, data and AI model training must satisfy additional non-functional requirements unrelated to cybersecurity (e.g., ensure that AI models match the computing power of drones). Finally, AI models and related data must be securely shared along the entire collaboration and circulation processes [6].

We classify drones missions in three settings of increasing complexity in terms of degree of collaboration and requirements to be enforced.

- **SOSM** (*Single Organization, Single Mission*): drones in a swarm are equipped with robust AI models (re)trained before the collaborative mission according to the mission peculiarities, and retrained according to data collected during the mission, upon mission completion.
- **SOMM** (*Single Organization, Multiple Missions*): drones in different swarms are equipped with robust AI models (re)trained *i)* before the mission according to its requirements, *ii)* according to a subset of data collected during different missions, upon their completion, considering the missions, data, and models peculiarities.
- **MOMM** (*Multiple Organizations, Multiple Missions*): drones in different swarms owned by different organizations belonging to the same federation (collaboratively or individually) perform their missions with robust AI models. Upon mission completion, robust AI models are retrained with data collected during the missions and across organizations, in a privacy-preserving fashion.

The *degree of trust* among drones, swarms, and organizations, varies across the three settings. Each setting introduces additional conflicting requirements, related to the need to re-use as much as possible existing artifacts (e.g., collected data) without opening the door to further threats and privacy violations (e.g., across organizations).

2.2 Use Cases

The settings in Sect. 2.1 are particularly relevant when the mission area is large, hostile, and difficult to cover by an individual organization or jurisdiction, as discussed in the following scenarios of military concern [14].

- **Collaborative mine detection.** Research has long been attempting to use automated detection and removal due to the extreme danger of the task. Recently, this issue has been approached using drones and AI techniques (e.g., [7]), but several challenges remain open (e.g., the presence of adversaries).
- **Collaborative border surveillance.** Border surveillance is traditionally based on physical barriers and human oversight. Drones can dramatically increase efficiency [13], though several challenges remain open (e.g., the security of the AI models used by the drones [12]).
- **Collaborative search and rescue.** Search and rescue missions are performed in response to disasters and large-scale accidents, as well as during mass migrations, to localize and rescue victims and people in danger. Drone-based solutions are emerging (e.g., [1]), but several challenges remain open (e.g., cross-organization and cross-border cooperation, and the lack thereof).

Our proposal will strengthen and fine-tune AI models in each aforementioned setting and use case.

3 Our Approach

We propose a solution that aims to improve the robustness of AI models in distributed and collaborative AIoT systems, with particular reference to drones swarms, where conflicting functional and non-functional requirements must be balanced. The proposed approach *i)* manages the secure evolution of AI models used by drone swarms and *ii)* collaboratively enhances their quality across swarm, mission, and organization boundaries. It implements the three-step process in Fig. 1 following the drone swarm life cycle, as follows.

- **(De)Briefing**: it supports robust AI model generation at Step *Preparation* according to the analysis of mission data (e.g., information on the drones, the mission to be executed, the drone and swarm behavior and activities during the execution of previous missions). Step (De)Briefing first analyzes the trustworthiness of the collected sample/data points [2]; it also evaluates the trustworthiness of the drones after mission completion. The output of this analysis is a *data trust score* associated with each sample/data point (Sect. 4). Each organization performs Step *(De)Briefing* independently.
- **Preparation**: it prepares robust AI models to be used in a mission. According to the setting (*SOSM*, *SOMM*, or *MOMM*), it (re)trains a (set of) robust AI model according to the *i)* functional and *ii)* non-functional requirements of the mission, *iii)* data trust score of each data point selected for (re)training. It balances conflicting requirements and trains AI models such that quality attributes fit the mission requirements at best and with an adequate level of robustness (e.g., [16]). For instance, in *MOMM*, Step Preparation may perform Federated Learning building the AI model for the mission task by exchanging the parameters of the local models in the FL protocol. As another example, Step Preparation may train a set of AI models to be dynamically

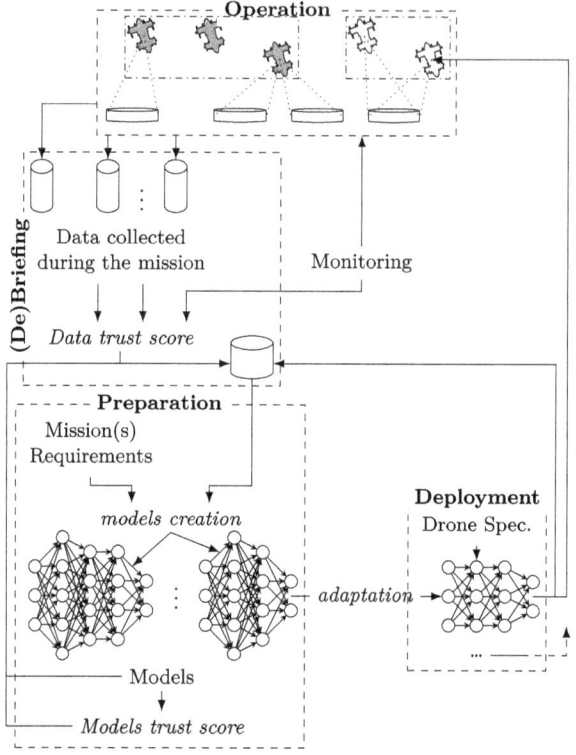

Fig. 1. Our approach.

selected during the missions (e.g., using dynamic classifier/ensemble selection techniques [9]). Each AI model and corresponding parameters are associated with a *model trust score*, as an aggregation of the data trust scores used to train the corresponding AI model (Sect. 4). Organizations perform Step Preparation independently (*SOSM*, *SOMM*) or collaboratively (*MOMM*) according to the setting.
- **Deployment**: it deploys the robust AI models created at Step Preparation on the swarms target of this proposal. Step Deployment adapts the trained AI model(s) to the specific drone where the model is deployed. This adaptation occurs along two main aspects of the drone: *i)* computing power, adopting footprint-reduction techniques (e.g., *ONNX Neural Compressor*[2]), *ii)* the peculiarities of the sensor(s), supporting preprocessing (e.g., feature extraction techniques). In a nutshell, AI models are adapted to the drone hardware and software. Specifically, Step Deployment packages each tailored AI model into a standard package (e.g., *ONNX* or *OpenXLA*[3]) ready to be physically

---
[2] https://github.com/onnx/neural-compressor.
[3] https://onnx.ai/, https://openxla.org/.

installed on drones. Organizations perform Step Deployment collaboratively but autonomously.

The approach in this abstract leverages the results achieved in the project *Prevention and detection of poisoning and adversarial Attacks on Machine Learning Models* (PALM) funded by TII, especially in Step (De)Briefing. PALM analyzed the impact of poisoning attacks [18] on ML and showed that the robustness of ML models against untargeted data poisoning can be significantly increased when a monolithic model (e.g., a random forest) is replaced by an ensemble of ML models trained on disjoint partitions of the dataset [2].

4 From Data Trustworthiness to Evolving Model Trustworthiness

The soundness and robustness of our approach depend on the ability to support collaborative and resilient model learning across multiple organizations. The latter requires a solution to trustworthy data circulation, where the trustworthiness of data collected and used by each drone is evaluated at the end of its mission, at different layers, as follows:

- *data layer:* the statistical distribution of the data collected by each drone is matched against a reference statistical distribution (prior knowledge) [16]. Anomalies can be reported for data trustworthiness evaluation;
- *behavior layer:* the robustness of a specific drone is evaluated by observing its behavior and configuration in operation, using assurance techniques [4,5]; for instance, sensors' configuration, the level of patches, the available resources, the flight itinerary, and connection downtime can affect the drone's robustness;
- *contextual layer:* the flight itinerary is used to retrieve the context where the specific drone operated identifying, for instance, high-risk areas and weather conditions affecting drone operation robustness and data trustworthiness.

According to these layers, a *data trust score* is associated with each data point collected during the missions and measures the corresponding trustworthiness. Data trust scores are retrieved during Step (De)Briefing, upon mission completion, and contribute to the evaluation of the model trustworthiness across continuous retraining. Data trust scores are used to build the training set for (re)training and calculate the *model trust score* measuring the trustworthiness of (re)trained models.

The *model trust score* is then calculated for each AI model at Step Preparation, when the datasets for (re)training are assembled and AI models built. We note that a global model trust score can be calculated on the basis of local model trust scores when a FL approach is used among the different organizations. We also note that data and model trust scores are continuously updated according to data and model life cycles, and their versions are immutably saved for future usage.

5 Conclusions

AI systems are increasingly characterized by scenarios where *i)* unknown and untrusted organizations carry out joint tasks using AIoT devices whose AI models are collaboratively trained, *ii)* conflicting functional and non-functional requirements insist on the same task and must be balanced, and *iii)* AI artifacts must safely circulate across organizations. In this context and with particular reference to drone swarms, we presented a novel approach to support secure and resilient evolution of AI models, which collaboratively enhances their quality across swarm, mission, and organization boundaries. Data and model trust scores are jointly used to support the trustworthiness of the mission(s) even in hostile environments. In particular, they are used to *i)* select the data for (re)training, *ii)* balance the different mission (non-)functional requirements, and *iii)* (re)train the AI models for the given mission(s).

Acknowledgments. Research supported, in parts, by *i)* TII under Grant 8434000394, *ii)* MUSA – Multilayered Urban Sustainability Action – project, funded by the European Union – NextGenerationEU, under the National Recovery and Resilience Plan (NRRP) Mission 4 Component 2 Investment Line 1.5: Strengthening of research structures and creation of R&D "innovation ecosystems", set up of "territorial leaders in R&D" (CUP G43C22001370007, Code ECS00000037), *iii)* project SERICS (PE00000014) under the NRRP MUR program funded by the EU – NextGenerationEU. Views and opinions expressed are however those of the authors only and do not necessarily reflect those of the European Union or the Italian MUR. Neither the European Union nor the Italian MUR can be held responsible for them.

References

1. Albanese, A., Sciancalepore, V., Costa-Pérez, X.: SARDO: an automated search-and-rescue drone-based solution for victims localization. IEEE Trans. Mob. Comput. **21**(9) (2022)
2. Anisetti, M., Ardagna, C.A., Balestrucci, A., Bena, N., Damiani, E., Yeun, C.Y.: On the robustness of ensemble-based machine learning against data poisoning. IEEE Trans. Sustain. Comput. **8**(4) (2023)
3. Anisetti, M., Berto, F., Banzi, M.: Orchestration of data-intensive pipeline in 5G-enabled Edge Continuum. In: Proc. of IEEE SERVICES 2022, Barcelona, Spain (2022)
4. Anisetti, M., Ardagna, C.A., Bena, N.: Continuous certification of non-functional properties across system changes. In: Proc. of ICSOC 2023, Rome, Italy, November–December 2023 (2023)
5. Anisetti, M., Ardagna, C.A., Bena, N., Bondaruc, R.: Towards an assurance framework for edge and IoT systems. In: Proc. of IEEE EDGE 2021, Guangzhou, China, December 2021 (2021)
6. de Arcaya, J.D., Torre-Bastida, A.I., Zárate, G., Miñón, R., Almeida, A.: A joint study of the challenges, opportunities, and roadmap of MLOps and AIOps: a systematic survey. ACM Comput. Surv. **56**(4) (2023)

7. Barnawi, A., et al.: A comprehensive review on landmine detection using deep learning techniques in 5G environment: open issues and challenges. Neural Comput. Appl. **34**(24) (2022)
8. Caruccio, L., Cirillo, S., Polese, G., Solimando, G., Sundaramurthy, S., Tortora, G.: Can ChatGPT provide intelligent diagnoses? A comparative study between predictive models and ChatGPT to define a new medical diagnostic bot. Expert Syst. Appl. **235** (2024)
9. Cruz, R.M.O., Sabourin, R., Cavalcanti, G.D.C.: Dynamic classifier selection: recent advances and perspectives. Inf. Fusion **41** (2018)
10. Di Nardo, E., Petrosino, A., Ullah, I.: EmoP3D: a brain like pyramidal deep neural network for emotion recognition. In: Proceedings of ECCV Workshops 2018, Munich, Germany, September 2018 (2018)
11. Hassija, V., et al.: Fast, reliable, and secure drone communication: a comprehensive survey. IEEE COMMST **23**(4) (2021)
12. Nguyen, K., et al.: The state of aerial surveillance: a survey. arXiv preprint arXiv:2201.03080 (2022)
13. Kim, S.J., Lim, G.J.: Drone-aided border surveillance with an electrification line battery charging system. JINT **92**(3) (2018)
14. Sawant, R., Singh, C., Shaikh, A., Aggarwal, A., Shahane, P., Harikrishnan, R.: Mine detection using a swarm of robots. In: Proceedings of ACCAI 2022. Chennai, India (2022)
15. Schlegel, M., Sattler, K.U.: Management of machine learning lifecycle artifacts: a survey. ACM SIGMOD Record (2023)
16. Singh, A.K., Blanco-Justicia, A., Domingo-Ferrer, J.: Fair detection of poisoning attacks in federated learning on non-I.I.D. data. KDD **37**(5) (2023)
17. Tian, J., Wang, B., Guo, R., Wang, Z., Cao, K., Wang, X.: Adversarial attacks and defenses for deep-learning-based unmanned aerial vehicles. IEEE IoT-J **9**(22) (2022)
18. Zhang, Z., et al.: Explainable data poison attacks on human emotion evaluation systems based on EEG signals. IEEE Access **11** (2023)

Open Access This chapter is licensed under the terms of the Creative Commons Attribution 4.0 International License (http://creativecommons.org/licenses/by/4.0/), which permits use, sharing, adaptation, distribution and reproduction in any medium or format, as long as you give appropriate credit to the original author(s) and the source, provide a link to the Creative Commons license and indicate if changes were made.

The images or other third party material in this chapter are included in the chapter's Creative Commons license, unless indicated otherwise in a credit line to the material. If material is not included in the chapter's Creative Commons license and your intended use is not permitted by statutory regulation or exceeds the permitted use, you will need to obtain permission directly from the copyright holder.

Adaptive Machines: Making Adaptive and Resilient Robots with Generative AI and Reinforcement Learning

Antoine Cully[✉] and the Adaptive and Intelligent Robotics Lab

Department of Computing, Imperial College London, London, UK
a.cully@imperial.ac.uk

Abstract. This short paper summarises some of the work presented in our Keynote on how Generative AI and Reinforcement learning can enable robots to face unforeseen situations like mechanical damage and autonomously adapt during their missions. In particular, we introduce a family of Generative AI called Quality-Diversity algorithms that are well-known for generating thousands of diverse and high-performing solutions to an optimization task. This diversity of solutions provides robots with an extensive set of alternative options to face unexpected situations. We also present how Quality-Diversity can be paired with Deep Reinforcement Learning to learn more complex policies, or with Generative Dynamics Models to ensure fast, safe, and continual learning and collection of data. Finally, we present how these tools can address Challenge 3: Adaptive Learning for Evolving Drone Operations.

Keywords: Learning algorithms · Damage Recovery · Robotics · Generative AI · Large Language Models

1 Introduction

Robots, such as drones and legged robots, hold immense potential to transform our societies by taking on the most dangerous tasks currently performed by humans. Today, thousands of lives are still at risk in mining operations, search and rescue missions, and disaster response efforts following natural catastrophes, such as earthquakes or nuclear incidents like the Fukushima disaster. To change this situation, robots must operate across a wide variety of environments, often under uncontrolled and unpredictable conditions. However, deploying complex robots in such challenging contexts increases the likelihood of damage or unexpected issues, such as a malfunctioning actuator or the loss of critical sensors. Moreover, robots are often tasked with critical missions where failure could have severe consequences, such as the loss of life in search and rescue missions. Therefore, it is essential to develop solutions that enhance the resilience of robots, enabling them to cope with unforeseen situations effectively.

Designing robust systems is a long-standing challenge in engineering. For example, significant advances have been made in aerospace engineering to ensure that airplanes remain safe across a wide range of scenarios. Traditional approaches to damage recovery in robotics typically involve two phases: self-diagnosis followed by the selection of pre-designed contingency plans [1]. While this approach has been successful in specific domains, such as aviation, scaling it to more complex environments, like exploring wreckage after a nuclear catastrophe, or to sophisticated robotic systems, such as swarms of drones, presents substantial challenges. At a certain point, it becomes impractical for engineers to anticipate every possible situation a robot might encounter, as well as all the potential sensor and actuator failures, to implement adequate diagnostic measures and pre-defined contingency plans.

2 Using Learning Algorithms to Enable Robots to Adapt to Unexpected Situations

An alternative approach for creating resilient robots involves the use of Machine Learning to enable robots to autonomously develop new behaviors in response to unforeseen circumstances. The Intelligent Trial and Error (ITE) algorithm [2] introduced by my group was among the first to demonstrate this capability, showing that a hexapod robot can adapt to unexpected mechanical damage, such as the loss of an entire leg, in under two minutes (see Fig. 1, and https://youtu.be/T-c17RKh3uE).

ITE achieves this rapid adaptation by combining the creative power of a generative algorithm known as Quality-Diversity (QD) [3] with the optimization capabilities of Bayesian Optimization [4]. QD algorithms are designed to generate a large collection of diverse and high-quality solutions. In the context of damage recovery for legged robots, QD generates thousands of distinct locomotion gaits, each utilizing different combinations of the robot's legs while maintaining high performance. The generative capability of the algorithm was used to autonomously generate more than 13,000 different gaits, each showing different locomotion properties (see https://youtu.be/IHQgnpSphEI). Although these gaits are initially designed for an intact robot, their diversity ensures that some will remain effective even if the robot sustains damage. For instance, if a leg is impaired, all gaits that do not rely on that leg will still function effectively. This computationally expensive step is run once before the deployment of the physical robot. Then, during deployment, ITE leverages this principle by using Bayesian Optimization—a data-efficient, black-box optimization algorithm—to identify the most appropriate gait for a given unexpected situation [4]. This step is computationally very efficient and can be executed on low-power devices like smartphones. The generative capabilities of QD are at the core of the resilience properties of ITE. It is interesting to note that this approach developed in robotics is now used also in the training of Large Language Models: Facebook uses Quality-Diversity to increase the robustness of their model LLAMA3.2 [5].

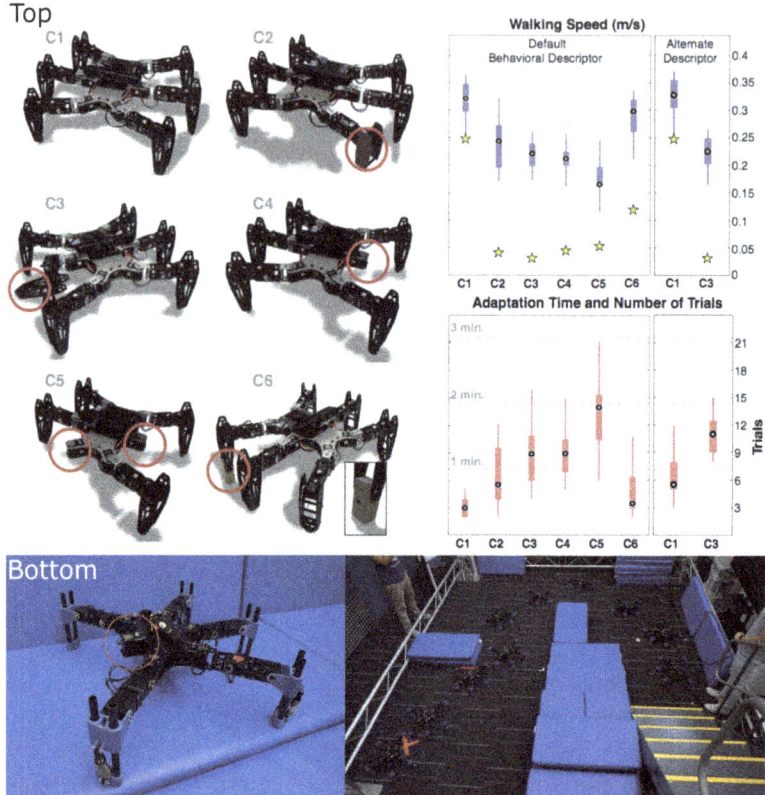

Fig. 1. (Top) Figure adapted from ITE [2] showing the recovered performance and adaptation time of a hexapod robot suffering from various mechanical damage situations. The black circles represent the median, the boxes refer to the interquartile difference, and the whiskers extend to the min/max values of the data. (Bottom) Figure adapted from HTE [6] showing a hexapod robot with a missing leg autonomously recovering from the situation and solving a maze navigation task.

3 Pushing the Frontiers of Resilient Machines

While the ITE algorithm has provided a compelling proof of concept that generative AI and learning algorithms can enhance the resilience of machines, several critical questions and technical challenges remain unresolved. For example, how can we ensure safe data collection during adaptation? How can we efficiently explore vast search spaces to discover innovative solutions? And how can we enable rapid learning of new behaviours directly on robots equipped with embedded systems? In the following sections, we present the solutions our research group has proposed to address these challenges.

3.1 Generating More Complex and Diverse Behaviours

While the original ITE algorithm was limited to small robotic controllers, producing only primitive movements, we have advanced this approach by integrating Quality-Diversity (QD) algorithms with Deep Reinforcement Learning (DRL). This integration leverages the creative potential of generative AI (through QD) and the data efficiency of gradient-based optimization (via DRL). Our research has led to the development of several algorithms that build on this synergy [7–10]. Notably, the DCG-MAP-Elites algorithm can generate thousands of complex robotic policies, each with over 20,000 parameters, that solve the same task in unique ways.

To further increase the diversity of behaviours that a robot can exhibit, we developed the Hierarchical Trial and Error (HTE) algorithm [6]. HTE composes complex skills by hierarchically combining modular components, resulting in a combinatorial expansion of the possible behaviours a robot can execute. This approach not only offers multiple ways to solve the same task but also enables the robot to tackle different tasks in various ways, thereby enhancing both its versatility and resilience. This is instrumental to recovering from damage while executing a complete mission (see Fig. 1 and https://youtu.be/vslztgdk-Zs).

We have also harnessed the power of Large Language Models (LLMs) to improve the training of robots in simulation. Our algorithm called OMNI-EPIC [11] uses LLMs to autonomously generate new training environments at the source code level (e.g., in Python), so we can challenge the generated policies more effectively. This combination of QD algorithms with LLMs creates an open-ended, never-ending list of challenges that our robots can practice in simulation before deployment [11]. This approach not only enriches the training process but also broadens the range of situations the robots can handle, which is crucial for enhancing their resilience.

3.2 Fast, Continuous and Safe Data Collection

Learning new behaviours during a robot's mission inherently carries risks. To ensure safe, continuous learning and data collection, we have proposed combining Quality-Diversity algorithms with Generative Dynamics Models (based on Transformer networks or diffusion networks) that learn the distribution of the consequences of the robot's actions [12,13]. This approach allows our QD algorithms to first simulate their next actions in "imagination," ensuring a sufficient level of safety before executing the action in the real world. This method has enabled our algorithms to maintain learning capabilities even in cluttered environments, successfully avoiding collisions with obstacles [13].

Moreover, the synergy between QD and Generative Dynamics Models offers an additional benefit: it allows the robot to discard not only unsafe actions but also uninformative ones. After simulating an action, the robot evaluates the potential information gain if the action were executed and selects only the most informative ones. This process enables our algorithm to run 20 times faster

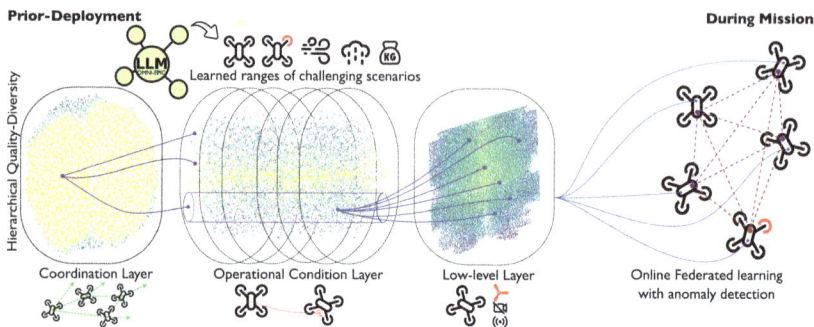

Fig. 2. Concept diagram of the proposed method to address Challenge 3: Adaptive Learning for Evolving Drone Operations, which combines our LLM OMNI-EPIC [11], Hierarchical QD[6], and our Federated Learning algorithm [14].

than competitive baselines by avoiding unnecessary real-world evaluations of actions [12].

To deploy these algorithms on robots, we implemented them using JAX [15], a Python framework that provides GPU acceleration, even on low-power embedded devices such as NVIDIA Jetson platforms [15,16]. This implementation was instrumental in our participation in the DARPA "Learning Introspective Control (LINC)" research challenge, where it enabled us to advance to the next phase of the ongoing program. The resulting algorithm FLAIR allowed a tracked robot to adapt to environmental changes and impairments in under 200 ms [17], such as sudden shifts in ground friction, damaged motors, and external forces like the Bernoulli effect in canals. This was achieved using exclusively embedded (noisy) sensors and low-power computational resources (Jetson AGX ORIN). This approach is designed to enhance the safety and reliability of logistics operations and vehicles across diverse conditions, potentially preventing incidents like the Ever Given container ship blockage in the Suez Canal (more: https://www.youtube.com/watch?v=D8Wbi_lcN4I).

4 Going it Further: Adaptive Learning for Evolving Drone Operations

The tools discussed in our keynote presentation, and summarized in this paper, represent an ideal suite for advancing drone operations through adaptive learning. While our previous research primarily focused on ground vehicles, including multi-legged and tracked robots, the proposed algorithms are inherently robot-agnostic. Their foundational concepts apply to any robotic platform, making drones a logical and promising next step. Moreover, our open-source implementation of these algorithms in JAX facilitates exceptionally fast execution on embedded or edge devices like the Jetson Orin Nano, often used on drones.

The specificity of drone swarms is that they include multiple agents and are likely to operate in an extremely large variety of situations. This causes a significant increase of scenarios that can affect the operations, from low-level mechanical damage, to large-scale situations requiring complete reorganization of the swarm. To address these challenges, our proposal (see Fig. 2) integrates three cutting-edge algorithms developed by our lab to: 1) leverage **Large Language Models (LLMs) to autonomously generate stress-testing scenarios** to train the adaptation algorithm, 2) employ **our hierarchical adaptation algorithm** to ensure resilience at all levels of a mission, from low-level mechanical damage to fleet-wide reconfiguration, and 3) utilize **Federated Learning** to accelerate adaptation while identifying drones experiencing individual anomalies.

The first contribution will harness OMNI-EPIC's environment generation capabilities [11] to create an extensive and open-ended collection of challenging scenarios. These scenarios will be used to rigorously test the adaptation algorithm before deployment, thereby maximizing its ability to handle a broad range of potential situations. Because OMNI-EPIC generates environments at the code level (e.g., python), it allows for the simulation of very low-level impairments, such as the loss of a propeller blade, sensor disturbances, communication disruptions within the swarm, and other critical situations. This diversity of scenarios not only promotes resilience but also prevents overfitting to a specific simulated environment.

The second contribution will use the Hierarchical Trial and Error (HTE) algorithm [6] to ensure resilience across all levels of the drone swarm. At the lowest level, drones will enhance their resilience to mechanical damage and sensor loss or corruption. At higher levels, the swarm's coordination will be fortified to withstand the loss of individual drones during missions, enabling rapid reorganization of the fleet. Intermediate levels will address higher-order impairments, such as environmental conditions like wind and limited visibility.

Finally, we will also extend the capabilities of Gaussian Processes with Collaborative Filters, a method we originally developed to rapidly adapt intelligent tutoring systems to new situations through Federated Learning [14]. This algorithm will be adapted for drones to allow each robot in the swarm to contribute to the collective learning process. Since all drones are likely to encounter similar environmental perturbations, sharing their experience will speed up the learning process. On the other hand, the algorithm is also able to isolate robots experiencing unique issues, such as mechanical damage or sensor faults, so that they do not disrupt the learning process for the rest of the swarm. Simultaneously, these drones will still benefit from shared knowledge relevant to the entire fleet.

These three contributions will enable drone swarms to operate effectively in missions where they encounter diverse, unforeseen challenges, continuously adapting their behaviour to enhance operational efficiency over time. While this is the core objective of Challenge 3: Adaptive Learning for Evolving Drone Operations, this approach also contributes to Challenge 1: Resilient Edge AI for Hierarchical Drone Swarms.

References

1. Verma, V., Gordon, G., Simmons, R., Thrun, S.: Real-time fault diagnosis [robot fault diagnosis]. IEEE Rob. Autom. Maga. **11**(2), 56–66 (2004)
2. Cully, A., Clune, J., Tarapore, D., Mouret, J.-B.: Robots that can adapt like animals. Nature **521**(7553), 503–507 (2015)
3. Chatzilygeroudis, K., Cully, A., Vassiliades, V., Mouret, J.-B.: Quality-diversity optimization: a novel branch of stochastic optimization. In: Black Box Optimization, Machine Learning, and No-Free Lunch Theorems, pp. 109–135. Springer, Heidelberg (2021)
4. Brochu, E., Cora, V.M., De Freitas, N.: A tutorial on bayesian optimization of expensive cost functions, with application to active user modeling and hierarchical reinforcement learning. arXiv preprint arXiv:1012.2599 (2010)
5. Samvelyan, M., et al.: Rainbow teaming: open-ended generation of diverse adversarial prompts. arXiv preprint arXiv:2402.16822 (2024)
6. Allard, M., Smith, S.C., Chatzilygeroudis, K., Lim, B., Cully, A.: Online damage recovery for physical robots with hierarchical quality-diversity. ACM Trans. Evol. Learn. **3**(2), 1–23 (2023)
7. Nilsson, O., Cully, A.: Policy gradient assisted map-elites. In: Proceedings of the Genetic and Evolutionary Computation Conference, pp. 866–875 (2021)
8. Pierrot, T., et al.: Diversity policy gradient for sample efficient quality-diversity optimization. In: Proceedings of the Genetic and Evolutionary Computation Conference (2022)
9. Faldor, M., Chalumeau, F., Flageat, M., Cully, A.: Map-elites with descriptor-conditioned gradients and archive distillation into a single policy. In: Proceedings of the Genetic and Evolutionary Computation Conference (2023)
10. Grillotti, L., Faldor, M., León, B.G., Cully, A.: Quality-diversity actor-critic: Learning high-performing and diverse behaviors via value and successor features critics. In: Forty-First International Conference on Machine Learning (2024)
11. Faldor, M., Zhang, J., Cully, A., Clune, J.: Omni-epic: Open-endedness via models of human notions of interestingness with environments programmed in code. arXiv preprint arXiv:2405.15568 (2024)
12. Lim, B., Grillotti, L., Bernasconi, L., Cully, A.: Dynamics-aware quality-diversity for efficient learning of skill repertoires. In: 2022 International Conference on Robotics and Automation (ICRA) (2022)
13. Lim, B., Reichenbach, A., Cully, A.: Learning to walk autonomously via reset-free quality-diversity. In: Proceedings of the Genetic and Evolutionary Computation Conference, pp. 86–94 (2022)
14. Cully, A., Demiris, Y.: Online knowledge level tracking with data-driven student models and collaborative filtering. IEEE Trans. Knowl. Data Eng. **32**(10), 2000–2013 (2019)
15. Chalumeau, F., et al.: Qdax: a library for quality-diversity and population-based algorithms with hardware acceleration. J. Mach. Learn. Res. **25**, 1–16 (2024)
16. Lim, B., Allard, M., Grillotti, L., Cully, A.: Accelerated quality-diversity through massive parallelism. Trans. Mach. Learn. Res. (2022)
17. Allard, M., Flageat, M., Lim, B., Cully, A.: Getting robots back on track: reconstituting control in unexpected situations with online learning. Under Review (2004)

Open Access This chapter is licensed under the terms of the Creative Commons Attribution 4.0 International License (http://creativecommons.org/licenses/by/4.0/), which permits use, sharing, adaptation, distribution and reproduction in any medium or format, as long as you give appropriate credit to the original author(s) and the source, provide a link to the Creative Commons license and indicate if changes were made.

The images or other third party material in this chapter are included in the chapter's Creative Commons license, unless indicated otherwise in a credit line to the material. If material is not included in the chapter's Creative Commons license and your intended use is not permitted by statutory regulation or exceeds the permitted use, you will need to obtain permission directly from the copyright holder.

Improving Resilience, Security, and Safety of Drones Through HTM-Based Adaptive Learning

Avi Mendelson, Leonid Azriel(✉), and Adam Ghadban

Technion, Israel, Institute of Technology, 32000 Haifa, Israel
leonida@technion.ac.il

Abstract. This proposal suggests significantly improving the safety, the security, and the reliability of drones via a novel design technique that combines fault tolerance (FT) design methodology, the use of Hierarchical Operations that employ a distributive version of Hierarchical Temporal Memory (HTM), and a novel set of design and testing tools. The system balances the need to guarantee no single point of failure and the requirement to operate at minimum resources such as area and power consumption. The system is designed to support a drone swarm, so, at the system level, a drone that loses a critical sensor, such as GPS, can retrieve the information from a neighbor drone. At the lower level of the hierarchy, an individual drone is controlled by a dedicated SoC that contains multiple RISC-V cores and a set of sensors. Each SoC contains an SMU (Security Management Unit), a small control unit that executes the top level of the distributed HTM algorithm and provides security services to the entire SoC. A set of dedicated sensors and targeted security counters are connected to the SMU; each sensor performs dedicated measurements and the lower level of the distributed HTM algorithm. The novel implementation of the algorithm is designed to consume minimum amount of power as long as the drone operates in a "normal" mode. Extra power is needed only if abnormal behavior is detected. The unique hierarchical implementation is designed to offer the best performance and power of the control system while providing a best-of-class recovery mechanism. Thus, it will enable an unstoppable system's operation even under extreme (unknown) conditions, malfunction of any single component (or at least all components that were designed with redundancy in mind), or security attacks. Validating the correct operation of FT real-time systems requires the development of new simulation and testing tools. This proposal suggests extending our current Gazebo-based simulation environment to emulate system-wide failures and security attacks on drone swarms together with a unique FPGA-based Fault injection tool. The targets for such operations will be determined from existing reported CWEs and by using a set of new tools, we are developing that take advantage of "information flow tracking" (IFT). The end goal of our proposal is to develop and demonstrate via an FPGA-based system, a proof of concept of the proposed method and its supporting tools.

Keywords: Security · Safety · Fault-Tolerance systems · HTM · Continuous operation

1 Problem Statement

To guarantee a reliable and continuous operation of real-time systems in general, and drones in particular, the system needs to address many potential hazards; such as mechanical and control malfunctions that could be caused by extreme environmental conditions, aging of mechanical and silicon modules, exposure of the system to extreme flight conditions, and up to potential security attacks. Current solutions can be divided between "special redundancy", such as the Triple Module Redundancy (TMR) that calls to duplicate each module three times and execute them in parallel, and time-based recovery that depends on the existence of an alternative execution path and use it if needed. The first option requires a significant number of resources and so is mainly suitable for expensive systems that can afford the use of many resources such as power. However, these systems are less practical for commodity usage and operation with a limited amount of resources. For such systems, we need to take a different approach; e.g., strive to get an indication that the system is about to fail soon, so we could start using alternative execution units before the failure affects the correctness of the system.

Thus, the use of time-based redundant systems heavily depends on the ability to detect hazards early enough, so the system could recover, or change its mode of operation ahead of time. To allow such a detection, while consuming minimum power, we propose a novel approach that is based on three "legs"; distributed implementation of the HTM algorithm that has already proven to be a preferred way for detecting malfunction and security hazards [1], an FT based design that aims to allow continuous operations even under hazards, and advanced testing and validation tools. Using all these techniques, enables the development of a new generation of reliable distributed real-time systems, and examine their characteristics under a realistic set of threats.

2 Background

This proposal includes a development of a new generation of HTM [2] algorithms. HTM is an unsupervised machine learning algorithm based on several known properties of the neocortex described in J. Hawkins's theory [3]. Figure 1 shows the essential components needed for HTM. Each time step, output streaming data is produced by the observed system. This data is transformed by the encoder into a sparse binary distributed representation (SDR) and used as an input to temporal memory (TM). SDRs are large binary vectors with small sparsity, meaning that only a small percentage of bits are '1'. This allows SDRs to have high representation capacity. For a typical vector size of 2048 and sparsity of 2%, SDR supports ~2.37×10^{84} different encodings. SDR representation allows high noise robustness. The probability of a false positive partial match of 50% with random SDR of the same size and sparsity is less than 4×10^{-26} [4].

Temporal Memory (TM) is a neural network organized as a collection of cell columns. Cells in a single column have the same forward input represented by bit 'i' in the input SDR. When the input SDR bit 'i' is '1', the feed-forward input signal to column 'i' is above the threshold, and the column is active. TM cells differ from the neuron model used in deep learning. In TM, in addition to the feed-forward input, each cell output can potentially connect to the lateral input segment of other cells in the network. In each step, the cell is in one of three states: active, off, and predicted. A cell is predicted if the input to one of the cell lateral segments is above a threshold. When a column is active, only predicted cells become active. If no cells are predicted, all column cells will become active (we call this column bursting). This allows the network to predict the next SDR input based on the current input SDR and the learned temporal context. The anomaly score can be naturally computed based on the prediction error by calculating the ratio of bursting columns to active columns.

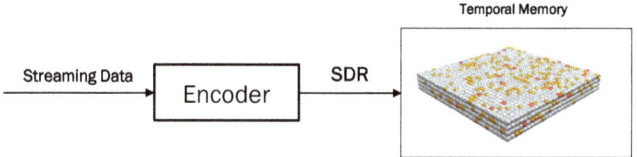

Fig. 1. Basic HTM Components

Experiments we conducted so far, using the SWAT data set **5** for anomaly detection, indicated that an HTM-based system can efficiently detect the creation of hazard conditions (Fig. 2) and can learn dynamically the creation of new attacks (Fig. 3).

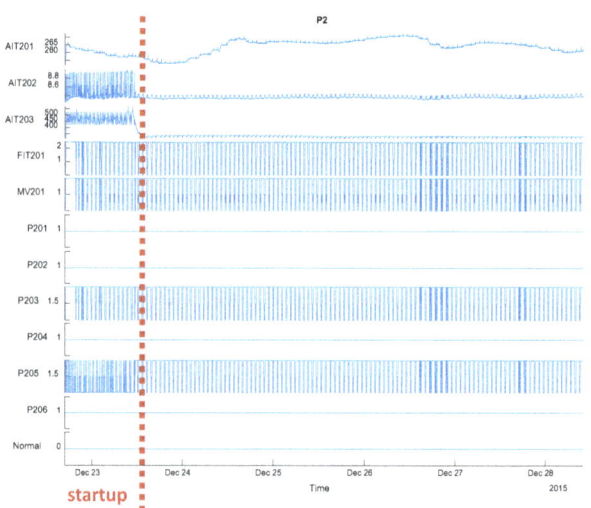

Fig. 2. Startup stage in the SWAT5 dataset.

So far, we focused on an efficient implementation of the HTM algorithm, in a "traditional manner", meaning by using a RISC-V processor which is equipped with a dedicated accelerator. Such an implementation is sufficient for tracking a single channel and it is suboptimal in terms of power and the execution time it requires. We suggest implementing a new generation of the algorithm – a distributive HTM that will be fully accelerated, be implemented in a distributed and hierarchical approach, by that being able to handle multiple channels simultaneously, could be reconfigurable at run-time to overcome potential failures and could be used to control distributed systems such as a drone swarm.

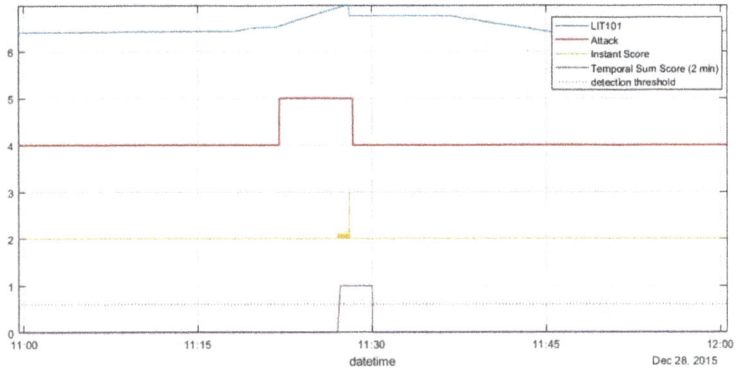

Fig. 3. Instant vs Temporal anomaly score for LIT101 channel

3 The Hierarchical Principle of Operation

We propose to use the hierarchical mode of operation as a leading principle for balancing between the need to use vast amounts of resources to maintain highly reliable systems and the need for efficient execution and low cost. Thus, we make the following assumptions: (1) The system is optimized for a "normal mode of operation" (2) at most a single point of failure (3) mixed critical tasks and enough resources to execute all critical tasks even under a single failure of any component (although, it can come at the expense of the performance of non-critical tasks).

At the Drone Swarm level, we assume a distributed control system that allows a single Drone to get services from the swarm. It can be used for saving power; e.g., not all drones need to access GPS devices most drones can get the needed location from neighbors, and this structure can be used as a recovery mechanism to recover from a device failure or to replace a device under a security attack.

At a single device level, we suggest building the SoC (System on a Chip) in a hierarchical structure, as indicated in Fig. 4.

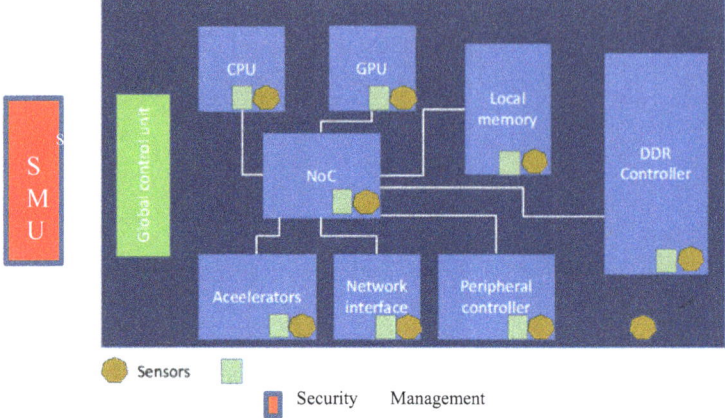

Fig. 4. Hierarchical structure of a control system

We assume that each SoC will have a highly secure SMU (Security Management Unit) that serves as an upper hierarchical control unit (we will extend its structure in the next section). The SMU will track the global state of the drone and will determine, how it should behave assuming "normal mode of operation". The SMU will also track information received from all sensors and in particular, new sequences that were never observed before and may indicate on failure or security attack. To perform this task, we are developing a new algorithm for HTM and termed it "distributed HTM" (DHTM) that splits the traditional HTM operations between the "upper" algorithm, which takes care of "global decisions" and may involve "rule-based" decisions, past statistics and white/black list of sequences, and "lower level" algorithm that aims to create sequences of events and make "simple decisions" such as if the current sequence can be considered as normal behavior or need to be reported to the upper level for farther manipulation.

An important role of the SMU (and the "upper algorithm") is to react to new sequences of events when detected. The system can "learn" the new sequence as a legal one, so it will accept it in future cases, or the system can self-heal itself by reconfiguring the hardware to avoid the impact of the fault or the security attack.

Note that most of the system's activity is done by the lower level of the HTM algorithm, and will be executed as part of the local control unit. This activity consumes very little power as long as the system runs in a normal mode of operation.

4 The SMU Protected Architecture

The SMU is a critical resource to guarantee the security of the drone's security. Thus, it must be built with sufficient protection against attacks and with the ability to recover from faults and attacks. We consider three classes of attacks/failures; fault injection (FI), side channel (SCA), and faults; e.g., aging of Silicon. To protect against them, SMU will incorporate a secure configuration of the IBEX core and other security-critical components.

To protect against the side-channel, we propose to build the SMU system using the share-based implementation approach. The share-based implementation is a cryptographic masking scheme that 'hides' the leakage of a secret via a side channel by splitting the secret into 'shares' using random masks and computing the result for each share separately [10, 11]. For example, splitting a secret value S to n additive shares $S = S_1 \oplus S_2 \oplus \cdots \oplus S_n$ makes choice of any number of shares i < n statistically independent of the secret. Such techniques are ubiquitous in the cryptography implementation, both hardware and software. In this project, we propose to make the general-purpose core (i.e. IBEX) side channel protected using implementations with shares so that any software that runs on this processor will be protected. This proposal is based on the research conducted in our group on the implementation of specific elements of a CPU (such as an ALU) using shares.

To protect the SMU against fault injection and other types of transient faults, we suggest building the SMU using double execution with local checkpoints. The use of double execution allows us to verify the correctness of the logic operations the SMU is executing, and the use of a local checkpointing mechanism enables (1) re-execution of operations in case of disagreement between the logic units and (2) in the case that the system does not agree after a given number of retries, to decide which core is the correct one (assuming single faulty core) and to continue the operations of the system.

To prevent the system from becoming a single point of failure, we suggest building it in a way that enables any of the RISC-V cores within the drone's SoC, to replace the main functionalities of the SMU.

5 Testing Toolbox

Testing secure and fault-tolerant systems is a complicated task since it requires generating specific faults at a given time and the system is built to mask the impact of such faults. A common way to test such systems is to build a cycle-accurate twin-simulator and inject faults into the system. The SCART 9 simulation we built, aims to allow the simulation of security attacks into a Gazibo-based drone flight simulator, but it is not sufficient for the simulation of a drone swarm and for using the actual FPGA model when available.

For these purposes, we are proposing to develop a set of additional tools;

1. Extending the existing SCART simulator to include drone swarm and the communication channels between them
2. Extending the FIJI-based tool 8 we developed, to include randomly generated faults, automatic check of correctness, and the ability to inject faults to multiple cards in a synchronized way

As an added tool to our debugging and test tool kit, we suggest developing a set of tools, based on Information Flow [12, 13], and in particular on Quantitative Information Flow (QIF) [14–16]. These tools aim to detect temporal and spatial locations for fault injection to maximize the probability of finding security problems in a given design. The Quantitative Information Flow (QIF) analysis is a relatively new field of research that measures the amount of information leakage rather than searching for a binary answer of whether such leakage exists. Based on probabilistic analysis and information theory, it may reduce the analysis complexity by using statistical models for the signals in the system. QIF can also be used for applications that look for a quantitative answer, such as the side channel leakage measurements.

Given a model, where the system is divided into secure (High) and non-secure (Low) domains, QIF can be used can be used to verify confidentiality (High to Low) and non-interference (Low to High) constraints. In the context of this work, we are interested in non-interference verification. By detecting points in the system that can affect the secure state, we can prebuild a set of weak links that should be verified for fault injection attack vulnerability.

6 Summary and Main Contributions

Building a real-time consumer-level fault-tolerant system is a major challenge since it requires to balance between contradicting demands; low cost and low power on one hand, but high safety and reliability on the other hand. So far, our research has mainly focused on mechanisms that aim to detect faults and security attacks. This proposal asks to extend the scope of our current research and to help to recover from such faults and security attacks while still achieving all the above demands; e,g, low power and smart consumption of resources. Thus, we propose the use of a hierarchical approach, together with early hazard detection that takes advantage of the DHTM mechanism. The combination of these mechanisms allows for sufficiently early detection of potential hazards so that the system can use time-based redundancy and avoid using costly TMR-based techniques.

We suggest building a novel control system for the drone swarm (AKA Al-Saqr-XX) based on the hierarchical principle of design that will allow the implementation of a real-time fault-tolerant system that consumes relatively low power. This proposal was mainly focused on the control unit of drones in general and drone swarms in particular. However, we believe that the same principle of operations can be applied to other critical real-time distributed systems. For such systems, we can offer an improved operational point; i.e., reduce cost while keeping the same (or even higher) level of resilience against faults and attacks. More than that, we can alter the active modules over time to extend the lifetime of the product and prevent aging and electromagnetic-related events, as we indicated in [6, 7].

References

1. Melitz, R., Mendeson, A.: The use of HTM and TSSE encoder for online anomaly detection. DCS-Water 2024 (2024)
2. Otahal, M., Keeney, D., McDougall, D.: HTM.core implementation of hierarchical temporal memory (2019). https://github.com/htm-community/htm.core/
3. George, D., Hawkins, J.: Towards a mathematical theory of cortical micro-circuits. PLoS Comput. Biol. **5**(10), e1000532 (2009)
4. Melitz, R., Mendelson, A.: Hierarchical online anomaly detection for cyber-physical systems using HTM with temporal variation encoder. submitted for publication to Int. J. Critical Infrastruct. Prot. (2023)

5. Secure Water Treatment (SWAT) Testbed. Technical Report: ITrust, centre for research in cyber security, Singapore university of technology and design (2020). https://itrust.sutd.edu.sg/itrust-labs-home/itrust-labs_swat/
6. Gabbay, F., Mendelson, A.: Electromigration-aware architecture for modern microprocessors. J. Low Power Electron. Appl. **13**(1), 7 (2023)
7. Gabbay, F., Ramadan, F., Ganaiem, M., Rosenthal, O., Bashari, L.: Effect of Asymmetric Transistor Aging on GPGPUs. In: Proceedings of the 5th International Conference on Microelectronic Devices and Technologies (MicDAT 2023), pp. 52–56 (2023)
8. Fibich, C., Tauner, S., Rössler, P., Horauer, M., Matschnig, M., Taucher, H.: Fiji: fault injection instrumenter. EURASIP J. Embed. Syst. **2019**(2) (2019). https://doi.org/10.1186/s13639-019-0088-7
9. Girstein, K., Rahimi, E., Mendelson, A.: SCART: simulation of cyber attacks for real-time. arXiv preprint arXiv:2304.03657 (2023)
10. Groß, H., Mangard, S., Korak, T.: Domain-oriented masking: compact masked hardware implementations with arbitrary protection order. Cryptology ePrint Archive (2016)
11. Nikova, S., Rechberger, C., Rijmen, V.: Threshold implementations against side-channel attacks and glitches. In: Ning, P., Qing, S., Li, N. (eds.) International Conference on Information and Communications Security. Springer, Heidelberg (2006). https://doi.org/10.1007/11935308_38
12. Coşkun, E.N.D., Ahmadi-Pour, S., Hassan, M., Drechsler, R.: Security coverage metrics for information flow at the system level. In: 2024 29th Asia and South Pacific Design Automation Conference (ASP-DAC), Incheon, Korea, Republic of 2024, pp. 945–950 (2024)
13. Reimann, L.M. et al.: QTFlow: quantitative timing-sensitive information flow for security-aware hardware design on RTL. In: 2024 International VLSI Symposium on Technology, Systems and Applications (VLSI TSA), HsinChu, Taiwan, pp. 1–4 (2024)
14. Hu, W., Ardeshiricham, A., Kastner, R.: Hardware information flow tracking. ACM Comput. Surv. (CSUR) **54**(4), 1–39 (2021)
15. Guo, X., Dutta, R.G., He, J., Tehranipoor, M.M.,. Jin, Y.: QIF-Verilog: quantitative information-flow based hardware description languages for pre-silicon security assessment. In: Proceedings of the 2019 IEEE International Symposium on Hardware Oriented Security and Trust, HOST 2019, pp. 91–100 (2019)
16. Reimann, L.M., Hanel, L., Sisejkovic, D., Merchant, F., Leupers, R.: QFlow: quantitative information flow for security-aware hardware design in verilog. In: 2021 IEEE 39th International Conference on Computer Design (ICCD), 2021, vol. 2021-Octob, pp. 603–607 (2021)

Open Access This chapter is licensed under the terms of the Creative Commons Attribution 4.0 International License (http://creativecommons.org/licenses/by/4.0/), which permits use, sharing, adaptation, distribution and reproduction in any medium or format, as long as you give appropriate credit to the original author(s) and the source, provide a link to the Creative Commons license and indicate if changes were made.

The images or other third party material in this chapter are included in the chapter's Creative Commons license, unless indicated otherwise in a credit line to the material. If material is not included in the chapter's Creative Commons license and your intended use is not permitted by statutory regulation or exceeds the permitted use, you will need to obtain permission directly from the copyright holder.

Robustness of Visual-Based Aerial Navigation to Real-World Adversarial Attacks

Yaniv Nemcovsky[1], Chaim Baskin[2], and Avi Mendelson[1(✉)]

[1] Department of Computer Science, Technion – Israel Institute of Technology, 3200003 Haifa Technion City, Israel
yanemcovsky@campus.technion.ac.il, mendlson@technion.ac.il
[2] School of Electrical and Computer Engineering, Ben-Gurion University of the Negev, David Ben-Gurion Blvd. 1, 8410501 Be'er Sheva, Israel
chaimbaskin@bgu.ac.il

Abstract. Imaging technologies are pivotal in the emerging aerial navigation ecosystem. However, these technologies are vulnerable to adversarial attacks. Current methods for enhancing the adversarial robustness of learned models are primarily based on adversarial training. However, such methods show limited success when applied to the high-resolution optical sensors used in aerial navigation. As the robustness of adversarially trained models is known to improve under attacks that closely resemble those seen during training, we suggest enhancing the models' robustness by focusing on highly realistic attacks. In this work, we aim to evaluate the susceptibility of navigation systems to such attacks and to enhance their corresponding robustness. We first intend to develop a realistic threat model and demonstrate the susceptibility of navigation systems to these attacks. We then aim to present a robust aerial navigation system by enhancing the systems' robustness via adversarial training incorporating the threat model.

Keywords: Visual-based Navigation · Aerial Navigation · Adversarial Attacks · Adversarial Training

1 Problem Statement and Key Finding

Adversarial perturbations were first discovered in the context of deep neural networks (DNNs), where the networks' gradients were used to produce small bounded-norm perturbations of the input that significantly altered their output [12]. Such attacks target the increase of the model's loss or the decrease of its accuracy and were shown to undermine the impressive performance of DNNs in multiple fields, e.g., image classification [3,5,9,12], object detection [2,16], real-life object recognition [1,15], autonomous navigation [8,14].

Imaging technologies used for aerial navigation are susceptible to adversarial attacks. This susceptibility was shown for optical, LIDAR (laser rangefinder), or

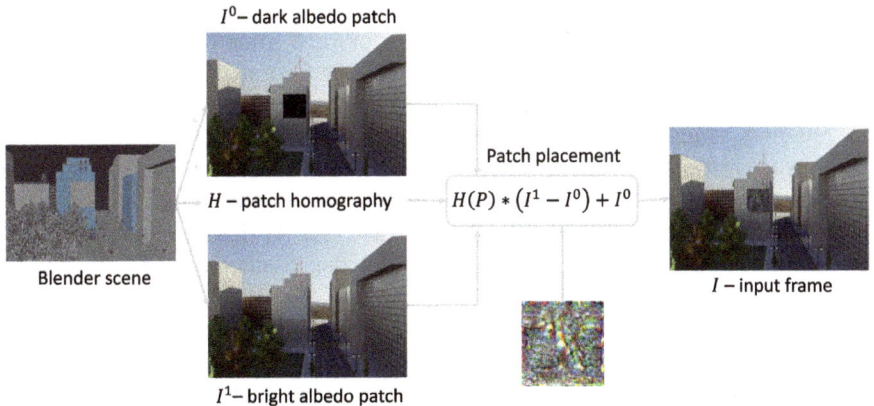

Fig. 1. Synthetic frame generation. The attack patch P is projected via the homography transformation H and is incorporated into the scene according to the albedo images I_0 and I_1.

Fig. 2. Real dataset frame generation. (a) Original image. (b+c) Black and white albedo approximations. (d) Adversarial patch projected onto the scene.

multiple imaging sensors working in unison[14]. Our previous work investigated the susceptibility of visual odometry models to universal adversarial perturbations over trajectories with numerous viewpoints. This work showed the feasibility of misleading a corresponding navigation system by disrupting its ability to position itself spatially in the scene [8]. These results were demonstrated for the highly realistic settings of a camera moving in a perturbed scene, where a physical passive patch adversarial attack is inserted into the scene and is perceived differently from multiple viewpoints. We have shown these results in both a simulated environment and real-world experiments, and we show the frameworks of the experiments in Fig. 1 and 2 accordingly.

The applicability of such attacks to real-world scenes represents a severe security risk as they could potentially push an autonomous system onto a collision course with some object by simply inserting a pre-optimized patch into a scene. The discussed attacks are highly relevant to drone swarm navigation settings as the swarm cameras provide multiple viewpoints of the scene during navigation. These viewpoints are then used to infer the swarm position and formation and to adjust the drones' trajectories accordingly. A patch adversarial attack could then induce deviations in the inferred position and formation and deviate the drones into a collision course with each other or with some object.

Adversarial attacks present vital security issues for autonomous systems and significantly threaten their proper conduct. Their demonstrated success suggests that hostile actions could be taken to harm the system, any passengers using it, and even bystanders. The need for adversarially robust autonomous systems is imminent, and non-robust systems cannot be genuinely autonomous as they require constant human attention to foil such attacks. At this moment, the future development of autonomous systems is highly dependent on their adversarial robustness.

Current methods of improving the adversarial robustness of learned models are primarily based on adversarial training, which incorporates adversarial attacks into the data during the models' training [3]. However, these methods show limited success when applied to high-resolution optical sensors such as those used in autonomous navigation [14]. Nonetheless, in real-world scenarios, the adversary is limited by access to the scene and the attacked system. In such scenarios, the adversary has limited knowledge of the position of the attacked system and must account for various viewpoints to produce effective attacks; i.e., the attacks cannot be tailored to a specific input and must be universal adversarial perturbations [7]. Therefore, the scope of potential attacks is limited, especially when considering realistic means of inserting the adversarial signal into the scene. For the robustness of real-world aerial navigation systems, we must only consider attacks from this realistic scope. Moreover, adversarial training methods perform better under attacks that closely resemble those encountered during the training. Models' robustness can, therefore, be improved by focusing solely on realistic attacks.

This project aims to present adversarially robust visual-based aerial navigation systems. To this end, we first intend to develop highly realistic adversarial attacks representing the threat model posed on such systems. These attacks will be used to research the susceptibility of existing systems, and we intend to study their applications in multiple settings. Moreover, we consider the effect of such attacks when applied on a single drone compared to a swarm. We then intend to incorporate the developed attacks in an adversarial training scheme to enhance the adversarial robustness of navigation systems.

2 Methodology

In our previous work [8], we showed that it is possible to disrupt the navigation of a system with minimal access to a scene. However, the full scope of potential real-world adversarial attacks and their potential effects is unclear. Therefore, we first aim to construct an ensemble of highly realistic attacks that better present the threat to navigation systems. As a result, we limit the adversary to only having access to the navigated system instead of directly accessing the attacked system. Our ensemble of attacks considers various means of inserting the adversarial signal into the scene and aims to represent the potential scope of real-world attacks.

In addition, we consider the effect of these attacks on swarms compared to a single drone. In a drone swarm navigation setting, multiple viewpoints of the same perturbation exist at each given moment. Therefore, when optimizing attacks aimed at swarms, we consider the position of numerous drones and the corresponding viewpoints. We employ this attack ensemble in adversarial training schemes, where we consider approaches to adversarial training that aim to be robust over multiple attacks.

2.1 Adversarial Attacks

To study the application of adversarial attacks under varying use cases, we build upon our previous passive patch adversarial attacks and consider a more realistic approach where the adversarial perturbation changes over time in repetition. We now present the passive patch attack setting discussed in [8] and extend the setting to dynamic repeating patch attacks.

Let $\mathcal{I} = [0,1]^{3 \times w \times h}$ be a normalized RGB image space, for some width w and height h, let $\{I_t\}_{t=0}^{L} \subset \mathcal{I}$ be a set of trajectory's images of length L with corresponding sets of black and white albedo images, denoted as $\{I_t^0\}_{t=0}^{L}, \{I_t^1\}_{t=0}^{L}$, let $A : (\mathcal{I} \times \mathcal{I}) \to \mathcal{I}$ be a perturbation inserting a patch image onto a given plane in the scene, and let $H(I, P)$ denote the linear homography transformation of a patch image P to viewpoint I. For a passive patch $P \in \mathcal{I}$, the perturbed trajectory is then achieved by applying the fixed perturbation individually on each image in the trajectory. Similarly, for a repeating dynamic patch with images $\{P_j\}_{j=0}^{T-1} \subset \mathcal{I}$, the perturbed trajectory is achieved by applying corresponding patch images on the trajectory frames. Formally:

$$I^P = A(I, P) \equiv H(I, P) * (I^1 - I^0) + I^0 \tag{1}$$

$$\{I_t^P\}_{t=0}^{L} \equiv \{A(I_t, P)\}_{t=0}^{L} \tag{2}$$

$$\{I_t^{\{P_j\}_{j=0}^{T-1}}\}_{t=0}^{L} \equiv \{A(I_t, P_{t\%T})\}_{t=0}^{L} \tag{3}$$

where $*$ denotes element-wise multiplication, and % denotes the modulus operator. Let $VO : (\mathcal{I} \times \mathcal{I}) \to (\mathbb{R}^3 \times so(3))$ be a monocular visual odometry (VO) model, and we extend its definition to trajectories over consecutive frames $VO(\{I_t\}_{t=0}^{L}) \equiv \{VO(I_t, I_{t+1})\}_{t=0}^{L-1}$. Given a set of trajectories $\{\{I_{i,t}\}_{t=0}^{L_i}\}_{i=0}^{N-1}$, with corresponding ground truth motions $\{\{\delta_{i,t}^{t+1}\}_{t=0}^{L_i-1}\}_{i=0}^{N-1}$, and a criterion over the trajectory motions ℓ, a universal passive patch adversarial attack $P_{passive}$ aims to maximize the sum of the criterion over the trajectories. Similarly, a dynamic repeating patch adversarial attack $P_{dynamic}$ aims to maximize the same criterion. However, in real-world scenes, we must account for frame offset and differing frame rates of the perturbations compared to the navigation system's camera. This uncertainty in the attack application can be described as an extension of the uncertainty over the input, and we extend the universal attack scheme with Monte Carlo sampling over the attack application. We denote the number of samples as M, where we sample frame rate ratio FR and frame offset FO

between the dynamic patch and attacked system. Formally:

$$P_t^m \equiv P_{(t \cdot FR_m + FO_m)\%T} \tag{4}$$

$$L_A(\{I_t\}_{t=0}^L, P_t^m) \equiv \ell(\{VO(A(I_t, P_t^m))\}_{t=0}^L, \{\delta_t^{t+1}\}_{t=0}^{L-1}) \tag{5}$$

$$P_{passive} \equiv \arg\max_{P \in \mathcal{I}} \sum_{i=0}^{N-1} L_A(\{I_{i,t}\}_{t=0}^{L_i}, P) \tag{6}$$

$$P_{dynamic} \equiv \arg\max_{P \in \mathcal{I}} \sum_{i=0}^{N-1} \sum_{m=0}^{M-1} L_A(\{I_{i,t}\}_{t=0}^{L_i}, P_t^m) \tag{7}$$

Using a changing perturbation presents a more extensive scope of adversarial attacks than fixed perturbation, improving the attack's performance. However, such a dynamic approach to attacks presents additional challenges. The attacks' applicability depends on the means of inserting the adversarial signal into the scene, which may require substantial preparation of the scene by the adversary. We consider two attacks based on dynamic perturbations that differ in their approach to inserting the adversarial signal into the scene; we discuss the advantages and disadvantages of each concerning performance and applicability. We now present the resulting real-world dynamic adversarial attacks and detail the key performance indicator (KPI) expected for each attack compared to the deviation benchmarks achieved by the passive patch adversarial attack:

- video patch adversarial attacks: Our first suggested dynamic attack uses a video as the adversarial perturbation instead of a single image. The video is then repeatedly projected on the designated patch instead of the fixed image. This method aims to improve the attack performance substantially by enabling the deployment of more complex attacks, e.g., the perturbation can be changed to falsely indicate the movement of a navigation system in any desired direction. However, projecting the patch's video via a projector will significantly diminish the adversarial signal in daylight. Therefore, the video patch attack requires projection on a designated screen to be applicable in the daytime. This entails significant preparation of the scene by the adversary, thereby limiting the attack's overall applicability. The KPI for this task is for the attack to achieve $2 - 3\times$ the deviation benchmarks.
- laser projection adversarial attacks: Another dynamic attack we consider is using laser projection, where we project a changing pattern on any surface without the signal being diminished in the daytime. This method aims to give a more applicable alternative to the video patch. However, the possible perturbations are now limited to those the projector can produce. We model the possible perturbations as sparse adversarial perturbations, limited in the L_0 norm, and base this attack on recent work by our group [9]. Such bounds are known to reduce the attacks' performance, and we expect the laser projection attacks' performance to be lower than the video patch attacks. Nonetheless, as the laser projector's position and direction can be adjusted to intercept incoming drones, we consider such attacks highly applicative. Moreover, the

laser's heat signature may also be relevant for fooling thermal-imaging-based models. The KPI for this task is for the attack to achieve 2× the deviation benchmarks.

The two suggested attacks aim to represent different use cases where the emphasis is either performance or applicability. The video patch attack seeks to maximize performance with limited applicability, and the laser projection limits the scope of attacks for increased applicability. We can consider alternating between our suggested attacks depending on the environmental conditions. In such a configuration, both attacks can be applied via a projector, where the video attack is used when the environmental conditions allow for its high-quality projection. Similarly, we can consider the L_0 bounds on the laser projection to represent the environmental conditions where we focus the perturbation's energy when the conditions require such. Another way to test the performance of our attacks in various settings is to use generative models to augment the trajectory images. In this case, we extend the attacks optimization dataset to represent multiple conditions and other changes in the scene. This enables testing attacks' generalization to various conditions over the same scenes.

2.2 Adversarial Training

Our suggested ensemble of attacks includes the three previously discussed attacks: "physical passive patch," "video patch," and "laser projection." Based on these attacks, we aim to employ adversarial training schemes. We consider two methods in doing so, differing in their approach to achieving robustness over the different attacks in the ensemble. Our first method suggests training a single model to be robust over all the attacks, and the other suggests training multiple models to be robust on different attacks. The models' ensemble approach has previously achieved better robustness than a single model in various settings, and we expect to achieve similar results under our proposed attack ensemble. However, such an approach requires several models and more computational resources than training and deploying a single model. Therefore, we intend to compare performance and computational overhead between the suggested methods. We now discuss our suggested adversarial training schemes and detail the KPI expected for each compared to the average base system deviation benchmark over the attack ensemble.

- Adversarial Training over Multiple Perturbations: A direct method to achieving robustness for the attacks in the ensemble is to employ an adversarial training scheme that aims at robustness to multiple perturbations. Such schemes have been proposed previously [6,13], and we intend to build upon them for training our robust model. The KPI for this task is for the robust model to achieve 0.5× the deviation benchmark.
- Adversarial training of models' ensemble: Another approach to achieving robustness over multiple attacks is to train independent models on each attack in the ensemble. During inference, we then aggregate the outputs from the

models to output a final result. Similar approaches to training models and output aggregation have been previously proposed [4,11], and we intend to build upon them for training and deploying our ensemble. Training or fine-tuning the resulting ensemble is out of the scope of the current work, as doing so requires substantially more computational resources and thereby limits the scalability of this solution. Nonetheless, previous work by our group suggests a fine-tuning scheme for such ensembles [10], and we consider such for future work. The KPI for this task is for the robust model to achieve $0.3\times$ the deviation benchmark.

3 Significance

This project is significant in two ways. Firstly, we intend to evaluate the robustness of aerial navigation systems to highly realistic adversarial attacks. Understanding the threat of such attacks is significant for developing and deploying autonomous systems. Currently, the scope of this threat is not well understood and poses an unknown risk for the proper conduct of aerial navigation systems and other autonomous vehicles. Secondly, we propose an approach for improving the robustness of aerial navigation systems to such attacks. Our approach is based on adversarial training over the previously mentioned realistic attacks and aims to be applicable in real-world scenarios. We suggest various attacks differing in functionalities and strive to improve the adversarial robustness of the navigation systems under these attacks and their like.

References

1. Brown, T.B., Mané, D., Roy, A., Abadi, M., Gilmer, J.: Adversarial patch. arXiv preprint arXiv:1712.09665 (2017)
2. Chen, S.T., Cornelius, C., Martin, J., Chau, D.H.: Shapeshifter: robust physical adversarial attack on faster R-CNN object detector. In: Machine Learning and Knowledge Discovery in Databases: European Conference, ECML PKDD 2018, Dublin, Ireland, 10–14 September 2018, Proceedings, Part I 18, pp. 52–68. Springer (2019)
3. Goodfellow, I.J., Shlens, J., Szegedy, C.: Explaining and harnessing adversarial examples. arXiv preprint arXiv:1412.6572 (2014)
4. Kariyappa, S., Qureshi, M.K.: Improving adversarial robustness of ensembles with diversity training. arXiv preprint arXiv:1901.09981 (2019)
5. Madry, A., Makelov, A., Schmidt, L., Tsipras, D., Vladu, A.: Towards deep learning models resistant to adversarial attacks. In: International Conference on Learning Representations (2018). https://openreview.net/forum?id=rJzIBfZAb
6. Maini, P., Wong, E., Kolter, Z.: Adversarial robustness against the union of multiple perturbation models. In: International Conference on Machine Learning, pp. 6640–6650. PMLR (2020)
7. Moosavi-Dezfooli, S.M., Fawzi, A., Fawzi, O., Frossard, P.: Universal adversarial perturbations. In: Proceedings of the IEEE Conference on Computer Vision and Pattern Recognition, pp. 1765–1773 (2017)

8. Nemcovsky, Y., Jacoby, M., Bronstein, A.M., Baskin, C.: Physical passive patch adversarial attacks on visual odometry systems. In: Proceedings of the Asian Conference on Computer Vision, pp. 1795–1811 (2022)
9. Nemcovsky, Y., Mendelson, A., Baskin, C.: Sparse patches adversarial attacks via extrapolating point-wise information. In: AdvML Workshop in conjunction with Neural Information Processing Systems (2024)
10. Nemcovsky, Y., et al.: Smoothed inference for adversarially-trained models. arXiv preprint arXiv:1911.07198 (2019)
11. Pang, T., Xu, K., Du, C., Chen, N., Zhu, J.: Improving adversarial robustness via promoting ensemble diversity. In: International Conference on Machine Learning, pp. 4970–4979. PMLR (2019)
12. Szegedy, C., et al.: Intriguing properties of neural networks. arXiv preprint arXiv:1312.6199 (2013)
13. Tramer, F., Boneh, D.: Adversarial training and robustness for multiple perturbations. Adv. Neural Inf. Process. Syst. **32** (2019)
14. Xiong, Z., Xu, H., Li, W., Cai, Z.: Multi-source adversarial sample attack on autonomous vehicles. IEEE Trans. Veh. Technol. **70**(3), 2822–2835 (2021)
15. Xu, K., et al.: Evading real-time person detectors by adversarial t-shirt. arXiv preprint arXiv:1910.11099 (2019)
16. Zolfi, A., Kravchik, M., Elovici, Y., Shabtai, A.: The translucent patch: a physical and universal attack on object detectors. In: Proceedings of the IEEE/CVF Conference on Computer Vision and Pattern Recognition, pp. 15232–15241 (2021)

Open Access This chapter is licensed under the terms of the Creative Commons Attribution 4.0 International License (http://creativecommons.org/licenses/by/4.0/), which permits use, sharing, adaptation, distribution and reproduction in any medium or format, as long as you give appropriate credit to the original author(s) and the source, provide a link to the Creative Commons license and indicate if changes were made.

The images or other third party material in this chapter are included in the chapter's Creative Commons license, unless indicated otherwise in a credit line to the material. If material is not included in the chapter's Creative Commons license and your intended use is not permitted by statutory regulation or exceeds the permitted use, you will need to obtain permission directly from the copyright holder.

Challenge 4: – Enhanced Communication and Active Protection Framework

Enhanced Security and Coordination Framework for UAV Swarms Using Heterogeneous Communication Networks

Daniel Bonilla Licea[✉], Giuseppe Silano, and Martin Saska

Faculty of Electrical Engineering, Department of Cybernetics, Czech Technical University
in Prague, 12135 Prague, Czech Republic
{bonildan,giuseppe.silano,martin.saska}@fel.cvut.cz

Abstract. Unmanned Aerial Vehicle (UAV) swarms offer remarkable capabilities across numerous fields, performing complex tasks with high efficiency and adaptability. However, safeguarding these swarms from cyber threats poses a significant challenge. This paper addresses *"Challenge 4: Enhanced Communication and Active Protection Framework"*. We aim to solve key objectives by introducing a comprehensive framework aimed at bolstering the security and coordination of UAV swarms. Our framework incorporates communications-aware trajectory planning, the use of heterogeneous communication networks, advanced physical layer security measures, and Artificial Intelligence (AI)-driven strategies for detecting and mitigating attacks. By combining Optical Camera Communications (OCC) with conventional Radio Frequency (RF) systems and utilizing Reinforcement Learning (RL) and Federated Learning (FL), the proposed framework provides a robust, efficient, and secure operational environment for UAV swarms.

Keywords: UAV Swarms · Communication Framework · Cybersecurity · AI

1 Introduction

Unmanned Aerial Vehicle (UAV) swarms have the capability to operate autonomously and collaboratively, offering significant advantages over single-drone operations. These benefits include enhanced coverage, redundancy, and resilience [1]. To achieve effective coordination and communication within UAV swarms, various architectures have been proposed, with hierarchical architectures being the most widely adopted [2]. In a hierarchical structure, drones are organized in tiers, where higher-tier drones oversee and coordinate the actions of lower-tier drones. This arrangement simplifies command and control, reduces decision-making complexity at the individual drone level, and enhances scalability. The hierarchical approach is particularly favored for its efficiency in managing large swarms and its robustness in maintaining operational coherence [2].

However, hierarchical architectures also present challenges [3]. A major issue is the reliance on Radio Frequency (RF) communication systems for most UAV interactions.

D. B. Licea and G. Silano—Authors contributed equally to this work.

© The Author(s) 2026
M. Andreoni and S. Thakkar (Eds.): GENZERO 2024,
Proceedings of 1st GENZERO Workshop, pp. 117–123, 2026.
https://doi.org/10.1007/978-981-95-1050-4_14

This makes these systems vulnerable to attacks such as RF jamming, which can disrupt communication or Global Navigation Satellite System (GNSS) signals, and GNSS spoofing, which can manipulate positional data. Additionally, UAV identity spoofing can lead to unauthorized control, while eavesdropping on communications can compromise sensitive information [4]. These vulnerabilities highlight the need for robust security measures to protect the integrity and reliability of UAV communication networks.

A traditional approach relying solely on cryptographic measures is insufficient to address the security challenges faced by UAV swarms and to maintain a secure and highly robust operation. To ensure swarm security, it is essential to employ Artificial Intelligence (AI)-based techniques for detecting such attacks and to integrate advanced cryptographic methods with physical layer security techniques. Additionally, heterogeneous communication networks that use both RF and optical communications provide are highly effective in increasing the security of UAV swarms. This approach leverages the large bandwidth provided by RF systems while benefiting from the immunity of optical systems to RF jamming and their significantly higher resistance to eavesdropping [5]. Moreover, communications-aware trajectory planning is a powerful tool for enhancing physical layer security in UAV communications [6]. This technique involves designing flight paths that minimize exposure to potential attacks and optimize the security of communication links.

In this paper, we address *"Challenge 4: Enhanced Communication and Active Protection Framework"*. We aim to solve the key objectives of deploying advanced encryption and authentication techniques to safeguard drone communications, utilizing AI to optimize communication strategies and enhance resilience against environmental and malicious disruptions, implementing anomaly detection algorithms to quickly identify and counteract communication threats in real time, and establishing proactive defenses to maintain the integrity and reliability of communications under adversarial conditions with the proposed framework. The proposed approach includes the following key components:

- *Communications-Aware Trajectory Planning:* We build upon our previously developed framework for communications-aware trajectory planning [7,8] and adapt it to the multi-agent scenario. The framework is designed to increase communication reliability using directional antennas that are less vulnerable to jamming;
- *Heterogeneous Communication Networks:* We employ heterogeneous communication networks that we developed, incorporating both RF communication systems and Optical Camera Communications (OCC) systems for UAVs [2].
- *Physical Layer Security Techniques*: We exploit advanced physical layer security techniques enhance the robustness of UAV-to-UAV communication links, increasing their resistance to potential threats [9,10].
- *AI for Attack Detection and Adaptive Response:* We utilize AI to improve the detection of attacks and rely on Reinforcement Learning (RL) to adapt the swarm's behavior to ever-changing and unforeseen situations.

2 Communication and Coordination Framework

In this section, we present our comprehensive communication and coordination framework for hierarchical UAV swarms. This framework operates using a fog-edge formation mechanism, integrating fog and edge computing concepts [11] to enhance the processing capabilities and communication efficiency within UAV swarms. This approach enables real-time data processing and low-latency communication. The framework is divided into four key components: *Advanced Encryption and Authentication*, *AI-Optimized Communication Strategies*, *Real-Time Anomaly Detection*, and *Proactive Defense Mechanisms*. The proposed framework is developed using ROS2 and tested in the Gazebo simulator. Figure 1 outlines the proposed communication and coordination framework.

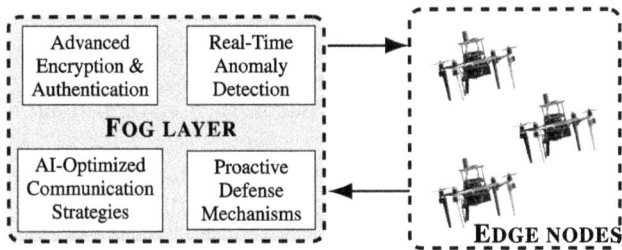

Fig. 1. Proposed communication and coordination framework. Arrows represent the data exchanged among drones.

2.1 Advanced Encryption and Authentication

One of the characteristics of OCC is its immunity to RF jamming, allowing many-to-many communications through the principle of spatial separability [6]. Additionally, eavesdropping on OCC is challenging without being in close proximity and equipped with the correct hardware. Heterogeneous communication networks that combine RF and optical communications offer the advantages of both systems. In this context, we propose a new authentication system for UAVs based on heterogeneous communications networks composed of OCC systems (e.g., UVDAR [2], designed by our team) and RF communications systems.

This authentication technique operates as follows: Each UAV is equipped with an OCC system composed of a camera and various LEDs that can emit arbitrary optical signals. Additionally, they are equipped with RF communications systems. Initially, UAV-zero (initially authenticated by the user) emits a command through its RF communication system for the other UAVs to initiate their authentication. The responding UAVs reply through their RF systems with a message that includes their ID and physical coordinates while simultaneously emitting valid optical signals through their LEDs. When UAV-zero receives the RF message, it uses its camera to verify the coordinates indicated by the other UAVs and retrieves the emitted optical signals.

If the initial message was emitted by a malicious entity that is not a UAV, UAV-zero would not detect any optical signals at the specified coordinates. If the malicious agent is a UAV equipped with the same OCC system, it would need to know the valid optical signals beforehand to be authenticated by UAV-zero. This requirement makes the authentication technique very difficult to deceive in practice.

2.2 AI-Optimized Communication Strategies

In this section, we introduce AI-optimized communication strategies that enhance the robustness and efficiency of UAV swarm operations using RL. RL is a type of machine learning where an agent learns to make decisions by performing actions in an environment to maximize cumulative rewards. For UAV swarms, RL agents are embedded within each UAV to optimize communication protocols in real-time.

Our RL framework includes the *state*, which represents the current conditions of the UAV swarm, including positions, signal strengths, and link statuses. The *action* space involves adjusting transmission power, selecting communication channels, and switching between RF and optical communication modes. The *reward* function balances communication reliability, latency, and energy consumption. Positive rewards are given for maintaining high-quality links, while negative rewards result from link failures or excessive energy use.

Each UAV, equipped with an RL agent, learns optimal communication strategies through interactions with the environment. Using algorithms like Deep Q-Networks, the agents continuously update their policies based on received rewards. This process ensures that the UAVs adapt to changing conditions and dynamic network topologies, enhancing the overall performance and resilience of the swarm.

AI-optimized communication strategies offer significant advantages. They allow UAVs to adapt to unforeseen challenges, such as interference or signal blockages, ensuring effective communication in various conditions. The decentralized nature of RL makes this approach scalable for large swarms. Additionally, continuous learning and adaptation improve the resilience of the swarm against communication disruptions, including RF jamming and signal interference. By dynamically adjusting protocols based on real-time conditions, these strategies enhance the reliability, efficiency, and security of UAV operations.

2.3 Real-Time Anomaly Detection

Detection of anomalies is crucial for identifying attacks and implementing appropriate countermeasures. Our approach leverages AI to detect abnormal behaviors and inconsistencies between different sensors that may indicate an attack. For instance, if an RF receiver detects an increase in received power along with an increase in the bit error rate, it can suspect a jamming attack. This information is then shared among all the UAVs via the optical network, which remains unaffected by RF jamming.

Once this data is distributed, a Federated Learning (FL) algorithm [12] processes the collective information from all UAVs to confirm the attack. FL enables UAVs to collaboratively learn a shared model while keeping their data decentralized, enhancing privacy and security. The algorithm can also attempt to estimate the location or direction of the attacker based on the shared data. This distributed AI approach not only detects jamming attacks but is also effective against more sophisticated threats such as proactive eavesdropping or deauthentication attacks, which rely on jamming.

2.4 Proactive Defense Mechanisms

The first line of defense is to hide communications from malicious agents. One way to achieve this is through the use of covert communications, where legitimate transmitters emit artificial noise when they do not have data to transmit. The idea behind this strategy is to confuse malicious nodes, making it difficult for them to determine when actual data exchanges are occurring among the UAVs.

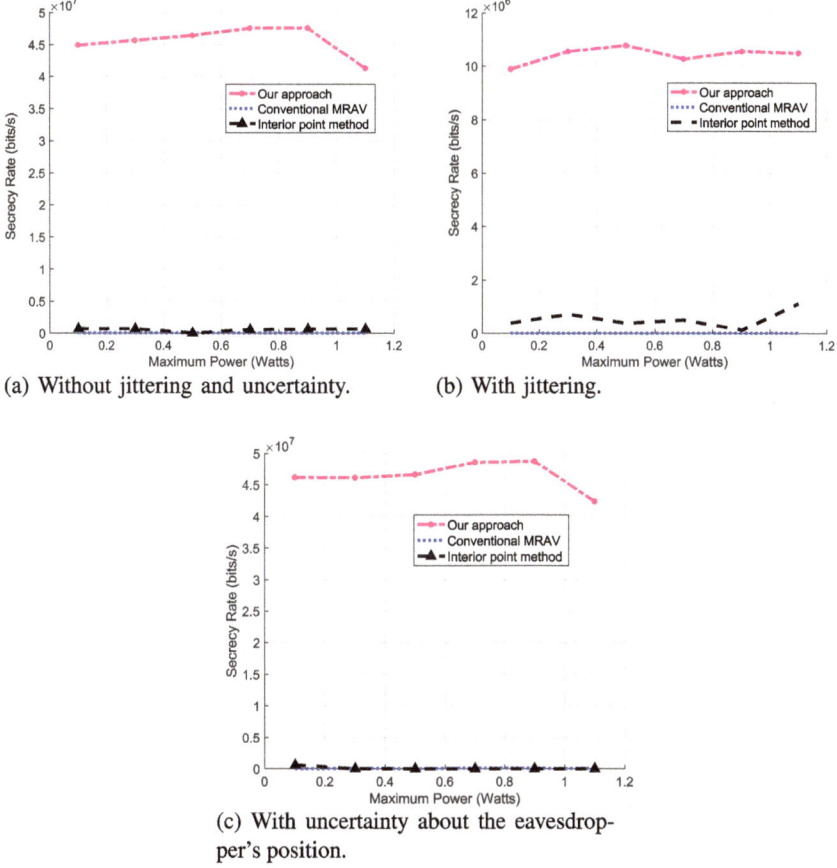

(a) Without jittering and uncertainty.

(b) With jittering.

(c) With uncertainty about the eavesdropper's position.

Fig. 2. The secrecy rate plotted against the maximum power for both the proposed approach and the two benchmarks without any jittering on the UAVs orientation and with perfect knowledge of the eavesdroppers position.

Another line of defense involves ensuring that legitimate transmitters focus their emitted power in the direction of legitimate receivers while minimizing power emission towards potential malicious receivers [10]. Similarly, this strategy aims to ensure that legitimate receivers experience high channel gain from legitimate transmitters and low channel gain from malicious transmitters [9].

This can be achieved through various methods. One solution is the use of multi-antenna arrays and beamforming techniques. Another approach involves using omni-directional UAVs and optimizing their orientation, as demonstrated in [9, 10], and depicted in Fig. 2. A third solution is to use under-actuated UAVs equipped with motorized antennas whose orientation can be controlled.

3 Conclusions

In this paper, we introduced a comprehensive communication and coordination framework to improve the security and operational efficiency of hierarchical UAV swarms. Our framework addresses key vulnerabilities in UAV communication systems, enhancing resilience against cyber threats such as jamming, spoofing, and eavesdropping. AI-driven methods for real-time anomaly detection and adaptive response ensure secure and efficient operations in dynamic and hostile environments. Future work will refine the AI models for more precise anomaly detection and explore integrating additional communication technologies to further enhance the system's robustness and scalability. Additionally, significant efforts will be dedicated to transitioning from simulation to real-world implementation, including conducting field experiments to validate the framework's effectiveness in practical scenarios and ensuring it performs reliably under real-world conditions.

Acknowledgment. This work was partially funded by EU under ROBOPROX (reg. no. CZ.02.01.01/00/22 008/0004590), by the Czech Science Foundation (GAČR) project no. 23-07517S, and by the CTU grant no. SGS23/177/OHK3/3T/13.

References

1. Chung, S.-J., et al.: A survey on aerial swarm robotics. IEEE Trans. Rob. **34**(4), 837–855 (2018)
2. Horyna, J., et al.: Decentralized swarms of unmanned aerial vehicles for search and rescue operations without explicit communication. Auton. Robot. **47**, 77–93 (2023)
3. Adil, M., et al.: A systematic survey: security threats to UAV-aided IoT applications, taxonomy, current challenges and requirements with future research directions. IEEE Trans. Intell. Transp. Syst. **24**(2), 1437–1455 (2023)
4. Kumari, N., et al.: Towards reliable identification and tracking of drones within a swarm. J. Intell. Robot. Syst. **110**(84), 1–31 (2024)
5. Chowdhury, M.Z., et al.: A comparative survey of optical wireless technologies: architectures and applications. IEEE Access **6**, 9819–9840 (2018)
6. Pandey, G.K., et al.: Security threats and mitigation techniques in UAV communications: a comprehensive survey. IEEE Access **10**, 112858–112897 (2022)
7. Bonilla Licea, D., et al.: When robotics meets wireless communications: an introductory tutorial. Proc. IEEE **112**(2), 140–177 (2024)
8. Bonilla Licea, D., et al.: Communications-aware robotics: challenges and opportunities. In: International Conference on Unmanned Aircraft Systems, pp. 366–371 (2023)

9. Bonilla Licea, D., et al.: Omnidirectional multi-rotor aerial vehicle pose optimization: a novel approach to physical layer security. In: IEEE International Conference on Acoustics, Speech and Signal Processing, pp. 9021–9025 (2024)
10. Bonilla Licea, D., et al.: Harnessing the potential of omnidirectional multi-rotor aerial vehicles in cooperative jamming against eavesdropping. In: IEEE Conference on Global Communications (2024, to Appear)
11. Pujol, V.C., et al.: Fog robotics-understanding the research challenges. IEEE Internet Comput. **25**(5), 10–17 (2021)
12. Li, L., et al.: A survey on federated learning. In: IEEE 16th International Conference on Control & Automation, pp. 791–796 (2020)

Open Access This chapter is licensed under the terms of the Creative Commons Attribution 4.0 International License (http://creativecommons.org/licenses/by/4.0/), which permits use, sharing, adaptation, distribution and reproduction in any medium or format, as long as you give appropriate credit to the original author(s) and the source, provide a link to the Creative Commons license and indicate if changes were made.

The images or other third party material in this chapter are included in the chapter's Creative Commons license, unless indicated otherwise in a credit line to the material. If material is not included in the chapter's Creative Commons license and your intended use is not permitted by statutory regulation or exceeds the permitted use, you will need to obtain permission directly from the copyright holder.

ZETInChat: Zero Trust Infrastructure with Dynamic Service Deployment via Chatbot in Mesh Networks

Guilherme Nunes Nasseh Barbosa and Diogo Menezes Ferrazani Mattos[✉]

LabGen/MídiaCom – PPGEET/TCE/TET, Universidade Federal Fluminense – UFF, Niterói, Brazil
{gbarbosa,menezes}@midiacom.uff.br

Abstract. The increasing frequency and severity of cyberattacks demand robust security solutions, especially for mission-critical environments. A key challenge is the dynamic configuration of secure services in Cloud Continuum for ad-hoc mesh networks, which traditional methods struggle to address. In this work, we propose ZETIn, a Zero Trust Infrastructure that leverages the power of Generative AI and a chatbot with LLMs to automatically configure services based on natural language inputs. The proposal ensures end-to-end security, simplifies the configuration process, reduces human error, and enhances network resilience and adaptability. Preliminary results demonstrate effective dynamic service configuration, high security standards, and improved network resilience, demonstrating that the integration of Generative AI with Zero Trust principles is a significant step forward in enhancing security and efficiency in mission-critical ad-hoc networks.

Keywords: Generative Artificial Intelligence · Zero Trust Infrastructure · Large Language Model · Mesh Network

1 Introduction

In recent years, cyberattacks have significantly increased in frequency and severity. In response to this trend, the *National Institute of Standards and Technology* (NIST) developed the *Zero Trust Architecture* [7,11]. The Zero Trust Architecture (ZTA) aims to provide a cybersecurity approach that minimizes implicit trust while requires continuous authentication, strict access control, and network micro-segmentation using Software Defined Perimeter (SDP) [8]. A key challenge is automatically creating secure services in Cloud Continuum environments for mission-critical Ad-Hoc Mesh Networks for autonomous systems, which are complex and error-prone to manage. These networks, essential for tasks like environmental monitoring and military operations, need flexible and secure configurations to ensure data resilience and integrity. Integrating wireless

---

This work was carried out with resources from the CNPq, FAPERJ, and CAPES.

devices adds complexity and vulnerability, underscoring the necessity of ZTA for continuous security.

In this work, we propose and demonstrate the ZETInChat, a Zero Trust platform that leverages Generative AI to automatically configure the network and services from natural language command descriptions. The system enables an interactive chatbot, integrated with large language models (LLMs), to receive commands and queries in natural language from operators and, based on these inputs, dynamically configure the necessary services on the mesh network. It includes applying security policies, allocating resources, and managing secure device communications. The platform uses advanced encryption and authentication techniques to ensure end-to-end security from the cloud to edge devices. This approach simplifies the configuration process and enhances the network's resilience and adaptability to different operational scenarios, improving incident response and protection against threats. The proposal aligns with the needs of autonomous systems, providing robust security and dynamic adaptability required for their operations.

Compared to existing solutions in the literature, our proposal offers a significant advantage by combining Zero Trust Architecture with Generative AI and mission-critical mesh networks. While many current approaches focus on static configurations or require manual adjustment intervention, our solution fully automates the process, reducing the risk of human error and speeding up the implementation of security measures. Additionally, integrating a chatbot with LLMs provides a more intuitive and accessible interface for operators, making network management easier even for those with less technical knowledge. Traditional solutions often fail to provide the flexibility and dynamics needed for mission-critical environments. Conversely, our proposal meets these demands, offering robust and adaptive security that adjusts to real-time needs.

2 Related Work

Previous works utilize Zero Trust Architecture concepts to tackle specific issues in companies and large organizations. Our proposal stands out because it integrates a Zero Trust platform with generative AI capabilities to configure services dynamically in mission-critical mesh networks.

Al-Hammuri *et al.* propose a Zero-trust-based scoring system to prevent medical errors in cloud-based healthcare information systems using machine learning and microservice-based authentication [1]. Our approach focuses on automatic network configuration and management based on uses OpenZiti and Ollama to implement a secure and flexible solution that processes natural language commands, facilitating automated and secure network and service configuration.

Kroculick focuses on theoretical development methodologies for Zero Trust architectures [6]. Our proposal, however, takes a more practical approach. We incorporate an LLM-based chatbot that allows fewer technical operators to dynamically configure complex networks, making the solution more accessible and applicable in real-world scenarios.

Tanimoto et al. address the scalability of software-defined perimeters (SDP) in diverse organizations by proposing models such as hierarchical and bridge [9]. In contrast, our project combines SDP with AI capabilities to not only scale but also adapt network configuration according to changes in the operational environment automatically.

Chandramouli and Butcher focus on access control in cloud-native applications using a service mesh and proxy infrastructure [3]. In contrast, our solution integrates automated service detection and configuration through natural language interactions, enhancing usability and security in Mesh networks.

While other works propose specific solutions for trust evaluation and network management[2,5], our approach is more comprehensive. We offer a holistic solution that not only evaluates trust but also dynamically configures and manages the network securely. By using OpenZiti and generative AI, we ensure a resilient and adaptable infrastructure for various mission-critical scenarios.

3 Zero Trust Architecture and Mesh Network

The proposed ZETIn platform relies on open-source solutions for deploying the Zero Trust Architecture (ZTA), the Mesh Network, and the Artificial Intelligence environment. OpenZiti[1] provides secure overlay networks, implementing ZTA with the Ziti Controller managing the Software Defined Perimeter (SDP) and Edge Routers acting as SDP Gateways. OpenZiti ensures strict access control, network traffic optimization, and scalability, supporting organizational expansion and robust IoT security [4,10]. The mesh network uses B.A.T.M.A.N.[2] (Better Approach to Mobile Ad-hoc Networking), a proactive routing protocol for Wireless Ad-hoc Mesh Networks, including MANETs. B.A.T.M.A.N. maintains information about accessible nodes, determining the best single-hop neighbor for each destination and facilitating efficient multi-hop routing[3]. The AI implementation leverages Open WebUI[4] and Ollama[5], creating a user-friendly and secure environment. Open WebUI, a self-hosted WebUI, supports various LLM runners, including Ollama and OpenAI-compatible APIs, ensuring data privacy and customization. Ollama enhances AI capabilities by processing natural language commands for dynamic network configurations. This integration provides a reliable and adaptable solution for managing network services, meeting the needs of modern, mission-critical environments.

4 The ZETInChat Proposal

Figure 1 illustrates the ZETin architecture for a robust, secure, and dynamic solution for configuring and managing mesh networks in mission-critical envi-

---

[1] https://openziti.io/.
[2] https://www.open-mesh.org/projects/open-mesh/wiki.
[3] Our proposal is agnostic to the B.A.T.M.A.N. version used, as long as IP connectivity between network nodes is ensured.
[4] https://openwebui.com/.
[5] https://ollama.com/.

ronments, utilizing Generative AI and Zero Trust principles. The architecture comprises two primary planes: the `Network Control Plane` and the `ZTA Plane`. Additionally, a `Forwarding Plane`, not shown in the figure, handles the actual data transmission.

At the Cloud layer of the `Network Control Plane`, the OpenZiti Controller is responsible for managing the Zero Trust security policy, ensuring continuous authentication and authorization of all access. The Generative AI Engine, a key component in this layer, processes natural language commands, enabling the dynamic configuration of network services. Ollama realizes the Generative AI Engine[6]. This dynamic configuration capability significantly enhances the network's adaptability and responsiveness. Additionally, an LLM-based chatbot provides a user-friendly interface for interaction, allowing users to issue commands and queries in natural language, which are then interpreted and executed by the AI engine. The LLM-based chatbot is deployed with OpenWebUI.

The Edge layer, a crucial part of the `Network Control Plane`, includes fog nodes that aggregate data from IoT devices and serve as intermediaries between the Cloud and the network's edge. These fog nodes provide local processing capabilities and enhance the network's scalability and responsiveness. Equally important in this layer are the OpenZiti edge routers, which play a key role in network security. They enforce security policies, ensuring that all communication is encrypted and that only authenticated and authorized devices can access network resources. This robust security measure significantly enhances the network's security. In the `ZTA Plane`, the mesh network leverages Batman-adv for efficient communication between devices.

Security within the mesh network is guaranteed by the Zero Trust architecture principles, which are enforced by the policies set by the OpenZiti Controller and executed by the edge routers. Users interact with the system through the chatbot interface, which runs in the Cloud and interprets natural language commands to dynamically configure the network and its services. This interaction allows for real-time adjustments and management of the network, enhancing its adaptability and robustness. The continuous application of security policies and the encryption of communication channels ensure that the network remains secure against unauthorized access and potential threats.

The architecture's design, encompassing the `Network Control Plane`, `ZTA Plane`, and the underlying `Forwarding Plane`, provides a comprehensive and robust solution for managing mission-critical Mesh networks. It combines the strengths of Generative AI and Zero Trust principles to deliver a secure, scalable, and highly responsive network environment, instilling confidence in its ability to meet the demands of mission-critical environments in mission-critical environments, utilizing Generative AI and Zero Trust principles. In the Cloud layer, the OpenZiti Controller manages the Zero Trust security policy, the Generative AI Engine processes natural language commands, and an LLM-based chatbot interacts with users. The Edge layer includes fog nodes that aggregate data from IoT devices and act as intermediaries, while OpenZiti edge routers enforce

---

[6] The proposal aims to leverage models such as LLaMA 3.1 and Falcon.

security policies. The Mesh network uses Batman-adv for efficient communication between IoT devices, with security ensured by the Zero Trust architecture. Users interact with the system through a chatbot interface that interprets natural language commands and dynamically configures the network and services. Security is ensured through encrypted communication channels and policies' enforcements.

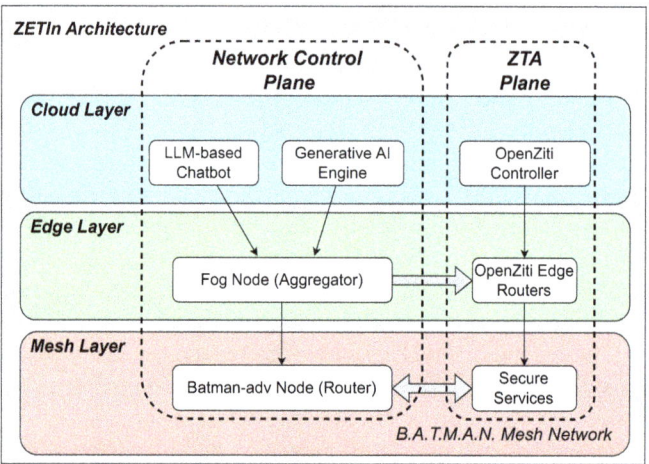

Fig. 1. The ZETIn architecture for managing mesh networks in mission-critical environments using Generative AI and Zero Trust. The Cloud layer includes an OpenZiti Controller, a Generative AI Engine, and an LLM-based chatbot for user interaction. The Edge layer has fog nodes and OpenZiti routers enforcing security. The mesh network uses B.A.T.M.A.N. for mesh communication, secured by Zero Trust. Users interact via a chatbot, which configures the network and services dynamically, ensuring encrypted communication and continuous security.

5 Demonstration

The demonstration will involve a mesh network composed of four Raspberry Pi devices, each acting as a node within the network. These devices will utilize the B.A.T.M.A.N. protocol for communication and routing. A gateway will connect the mesh network to the Cloud, where the OpenZiti Controller, Generative AI Engine, and LLM-based chatbot modules are deployed. The demonstration will showcase the dynamic configuration of network services through natural language commands processed by the chatbot, highlighting real-time adjustments and security policy enforcement. This setup will illustrate the seamless integration between the edge devices and the Cloud, ensuring secure, adaptable, and efficient management of the Mesh network in a mission-critical environment. The prototype is currently at Technology Readiness Level (TRL) 4, progressing towards TRL 5[7], as it has been validated in a lab environment and is now ready to be tested for performance and reliability in an operational setting.

6 Conclusion

The proposed ZETInChat platform integrates Generative AI with Zero Trust Architecture to address the complex challenges of dynamic service configuration in mission-critical mesh networks for autonomous systems. By leveraging OpenZiti and Ollama, the system ensures end-to-end security, simplifies configuration processes, reduces human error, and enhances network resilience and adaptability. Compared to existing solutions, our approach stands out by offering automated, real-time configuration via a user-friendly chatbot interface, ensuring continuous security and efficient management of network services. The demonstrated effectiveness in preliminary results, TLR4, underscores the potential of combining Generative AI and Zero Trust principles to significantly enhance the security and efficiency of mission-critical ad-hoc mesh networks. The proposal is in line to be adopted by autonomous systems, providing them with robust security and dynamic adaptability required for autonomous operations. Future work focuses on improving scalability for larger network environments, integrating advanced threat detection mechanisms using machine learning. Additionally, we envision analyzing the potential vulnerabilities of using LLMs, such as adversarial attacks and natural language processing errors.

References

1. Al-hammuri, K., Gebali, F., Kanan, A.: ZTCloudGuard: zero trust context-aware access management framework to avoid medical errors in the era of generative AI and cloud-based health information ecosystems. AI 5(3), 1111–1131 (2024). https://www.mdpi.com/2673-2688/5/3/55
2. Alboqmi, R., Jahan, S., Gamble, R.F.: A runtime trust evaluation mechanism in the service mesh architecture. In: 2023 10th International Conference on Future Internet of Things and Cloud (FiCloud), pp. 242–249 (2023)
3. Chandramouli, R., Butcher, Z.: A zero trust architecture model for access control in cloud-native applications in multi-cloud environments. Technical report, National Institute of Standards and Technology (2023)
4. Diaz Rivera, J.J., Khan, T.A., Akbar, W., Muhammad, A., Song, W.C.: ZT&T: secure blockchain-based tokens for service session management in zero trust networks. In: 2022 6th Cyber Security in Networking Conference (CSNet), pp. 1–7 (2022)
5. Khowaja, S.A., Nkenyereye, L., Khowaja, P., Dev, K., Niyato, D.: Slip: self-supervised learning based model inversion and poisoning detection-based zero-trust systems for vehicular networks. IEEE Wirel. Commun. **31**(2), 50–57 (2024)

---

[7] The demonstration will involve field testing to validate reliability, and expanding experiments with more nodes to assess scalability and performance in complex environments.

6. Kroculick, J.B.: Zero trust decision analysis for next generation networks. In: Disruptive Technologies in Information Sciences VIII, vol. 13058, pp. 278–286. SPIE (2024)
7. Sheikh, N., Pawar, M., Lawrence, V.: Zero trust using network micro segmentation. In: IEEE INFOCOM 2021 - IEEE Conference on Computer Communications Workshops (INFOCOM WKSHPS), pp. 1–6 (2021)
8. Syed, N.F., Shah, S.W., Shaghaghi, A., Anwar, A., Baig, Z., Doss, R.: Zero trust architecture (ZTA): a comprehensive survey. IEEE Access **10**, 57143–57179 (2022)
9. Tanimoto, S., Yangchen, P., Sato, H., Kanai, A.: Suitable scalability management model for software-defined perimeter based on zero-trust model. Int. J. Serv. Knowl. Manage. **7**(1) (2023)
10. Teerakanok, S., Uehara, T., Inomata, A.: Migrating to zero trust architecture: reviews and challenges. Secur. Commun. Netw. **2021**(1), 9947347 (2021). https://onlinelibrary.wiley.com/doi/abs/10.1155/2021/9947347
11. Zivi, A., Doerr, C.: Adding zero trust in BYOD environments through network inspection. In: 2022 IEEE Conference on Communications and Network Security (CNS), pp. 1–6 (2022)

Open Access This chapter is licensed under the terms of the Creative Commons Attribution 4.0 International License (http://creativecommons.org/licenses/by/4.0/), which permits use, sharing, adaptation, distribution and reproduction in any medium or format, as long as you give appropriate credit to the original author(s) and the source, provide a link to the Creative Commons license and indicate if changes were made.

The images or other third party material in this chapter are included in the chapter's Creative Commons license, unless indicated otherwise in a credit line to the material. If material is not included in the chapter's Creative Commons license and your intended use is not permitted by statutory regulation or exceeds the permitted use, you will need to obtain permission directly from the copyright holder.

Securing AI with AI: Novel Framework for Drone Communication Security

Andrea Bastoni[1], Rodolfo Pellizzoni[2], Miguel Costa[3,4](✉), Emanuele Parisi[5], Francesco Barchi[5], Andrea Acquaviva[5], and Sandro Pinto[3,4]

[1] Minerva Systems, Modena, Italy
andrea.bastoni@minervasys.tech
[2] University of Waterloo, Waterloo, Canada
rpellizz@uwaterloo.ca
[3] Universidade do Minho, Braga, Portugal
{miguel.costa,sandro.pinto}@dei.uminho.pt
[4] Zero-Day Labs, Braga, Portugal
[5] University of Bologna, Bologna, Italy
{emanuele.parisi,francesco.barchi,andrea.acquaviva}@unibo.it

Abstract. Autonomous drone swarms operating in potentially hostile environments and communicating over inherently insecure wireless channels require robust security architectures. While AI-based algorithms are effective in detecting communication anomalies and intrusions, their deployment in low-power environments like drones is challenging, and the integrity of AI-driven decisions can also be compromised. This paper discusses a comprehensive drone communication framework that enhances security by leveraging an efficient PMU design for minimally intrusive, real-time system tracing. Our approach improves system resilience by combining PMU data to (i) secure the integrity of AI-based decisions and (ii) detect intrusions through network profiling. The framework integrates the hardware Root-of-Trust of the target System-on-Chip to ensure integrity and privacy.

Keywords: Intrusion detection · Data and Communication integrity · Advanced PMU · AI-based NIDS

1 Introduction and Motivation

The rapid advancement of drone and Artificial Intelligence (AI) technologies is fostering innovation in a myriad of sectors, including agriculture, logistics, and military. One of the most promising developments in this field is the utilization of hierarchical drone swarms, which operate as coordinated units to achieve complex tasks more efficiently than individual drones. Given that these systems handle critical data in hostile environments and communicate using wireless channels, not secure by default, drone cyber-security is a major concern.

© The Author(s) 2026
M. Andreoni and S. Thakkar (Eds.): GENZERO 2024,
Proceedings of 1st GENZERO Workshop, pp. 131–137, 2026.
https://doi.org/10.1007/978-981-95-1050-4_16

Recent research has shown that AI is useful not only for route optimization and tailoring a drone for a given task, but also for detecting anomalies in drone communication [1] and preventing attacks by inspecting packet payloads [2]. Although AI-based algorithms have demonstrated remarkable efficacy in other security domains, such as malware detection, they are vulnerable to adversarial examples that compromise the decision integrity [3]. Therefore, while AI-based security can provide an additional protective layer, attackers can still undermine drone communication and functionality by subverting the integrity of AI.

In this work, we propose a comprehensive drone communication framework that takes a step further in drone security. Firstly, our framework comprises a novel technique to secure the integrity of AI-based decisions. We aim at leveraging the performance monitoring unit (PMU) of the main application processor (APU) to detect adversarial attacks against AI algorithms. Adversarial attacks are known [4] to generate input samples that activate the neurons of an artificial neural network (ANN) differently from legitimate inputs. Based on this premise, we investigate whether the PMU's fingerprint from adversarial examples differs from that of legitimate samples and whether this difference is sufficient to develop a robust defense.

Secondly, to enhance the security and ensure the integrity of critical AI algorithms for communication inspection, we leverage new System-on-Chip architectures equipped with a hardware Root-of-Trust, such as OpenTitan (OT) [5]. By executing the AI-based Network Intrusion Detection System (NIDS) within OT, we further capitalize on its capabilities. OT has proven effective as a cryptographic accelerator [6]. Here, we extend its utilization by performing on-the-fly decryption and analysis for deep learning intrusion detection. This method conducts AI-based network packet inspection immediately after decryption within OT, thereby preserving both NIDS integrity and packet privacy. Additionally, the AI-NIDS will utilize PMU information to profile network stack execution and identify anomalous behaviors in conjunction with payload inspection.

Our framework enhances the hardware-based tracing and monitoring that can be leveraged (e.g., [7]) to detect violations and intrusions on a critical systems. Specifically, we target the challenges of devising advanced PMUs to efficiently monitor the execution of applications such as AI-NIDS and AI-based algorithms. The collected traces are used to enhance the training and capabilities of AI-based integrity monitors and detection algorithms, which help detect malicious intrusions and improve system resilience. We aim for a minimally invasive and zero-overhead approach that enables a seamless, real-time tracing of the system. Our strategy can detect small deviations from standard application behavior and quickly react to threats affecting communication channels and system integrity. Given the integration in OT, the framework aligns well with advanced encryption protocols, while preserving NIDS and privacy in the communication.

2 Background and Related Work

In drone technology, adversarial attacks have been mostly used to induce navigation errors. By exploiting the wireless connection upon which drones rely, an attacker can potentially manipulate data collected from LIDAR, cameras, or GPS. This manipulation can cause the drone to deviate from its intended route or misinterpret its surroundings, potentially resulting in collisions [8]. Simulated experiments have shown that this type of attack could decrease the ability of a drone to detect an object in more than 60% of the cases [9]. Furthermore, by employing similar strategies, attackers can compromise the mission of a drone. As drones are widely used for surveillance and maintenance, manipulating data collected by, for example, cameras, can lead to erroneous decisions, often only identifiable through human surveillance, which renders useless the autonomous capability of drones. Recent studies have begun implementing AI to secure wireless communication [10], reducing the success of the attacks previously described. Also, inspecting network packets using deep learning approaches can be used to detect code reuse attacks such as Return-Oriented Programming (ROP), improving detection accuracy and reducing runtime overhead with respect to traditional methods [2]. These AI systems are also vulnerable to adversarial examples (e.g., [3,8]), and protecting their decision integrity remains an open challenge.

Defenses such as adversarial training have been extensively explored (e.g., [11]) to smooth the effect of adversarial examples on AI decisions. However, we have shown that these defenses can still be subverted [3]. Considering that adversarial examples activate the neurons of a neural network in a very different way from legitimate samples [4], in this work, we follow a disruptive approach and explore the feasibility of using the PMU to detect adversarial examples.

Detecting violations and intrusions in real-time involves analyzing applications' control flow, data flow, and events [4,7]. Leveraging hardware-based information for such detection can provide efficient and effective monitoring solutions [12,13], but often entails significant overheads that can impact system performance and reliability [13]. Our recent work on MemPol [14] demonstrates the benefits of exploiting additional hardware units to achieve fine-granular monitoring with minimal overheads. This technique offers an independent view of software execution across all levels, including hypervisors, operating systems, and applications.

To tackle these challenges, we plan to leverage and extend the work done by the team in the context of the AlSaqr and ZeroTrust initiatives. The target drone platform employs a multicore RISC-V architecture with advanced virtualization capabilities, running the Bao hypervisor [15], NuttX RTOS, and PX4 flight controller. It includes a hardware Root-of-Trust (RoT) based on OT, supporting secure boot and cryptographic operations (as implemented in AlSaqr project [6]); the secure enclave has access to a low-power RISC-V cluster optimized for AI operations. In ZeroTrust, we port deep learning algorithms for intrusion detection (specifically Control Flow Integrity) in this secure enclave. Core-level monitoring capabilities are being augmented through a snooper capable of collecting control-flow traces from the core and feed them to the RoT to enable machine learning-based integrity analysis. We further designed a system-level monitoring infrastructure consisting of a set of EVent Units (EVUs), integrated with the monitored components, which send micro-architectural events to a central Performance Monitoring Unit (PMU). The PMU includes a low-power instruction processor running specialized event-processing functions. In its current form, EVUs monitor memory transactions issued by application cores and Last-Level Cache (LLC).

We plan to exploit these features by performing NIDS integrated with packet decryption and utilizing PMU information to enhance NIDS accuracy through the analysis of network stack behavior.

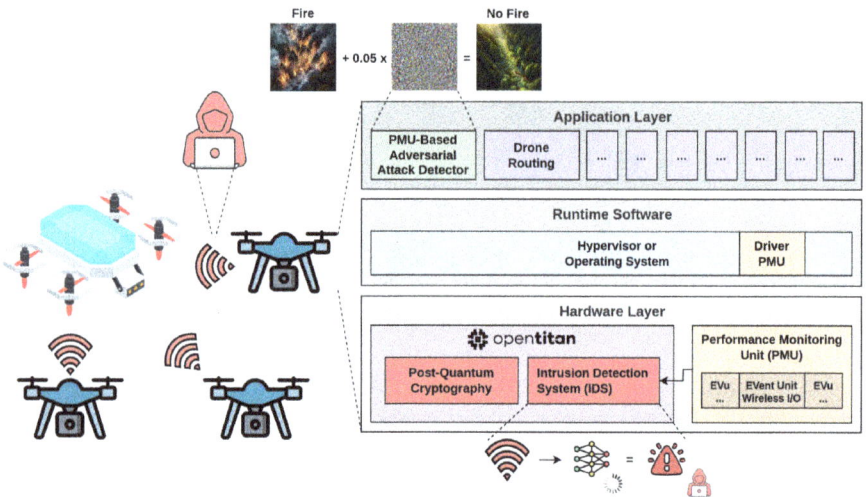

Fig. 1. System Architecture for an example surveillance drone swarm. Compromising communications, attackers could prevent the system to correctly detect a fire. Our framework leverage hardware PMU to detect attacks and intrusions at both application level and within the encrypted network communication.

3 Proposed Approach

The goal of our monitoring infrastructure is to increase observability into the system by aggregating micro-architectural information generated by distributed components across the platform. In line with [14], our proposed architecture centralizes the collection of event counters into an advanced PMU. By using such a dedicated hardware unit, our design provides improved access latency, better synchronization across events generated by disparate hardware components, and a unified programming model. Our design is open and based on common protocols such as AXI4; therefore, while proven on a RISC-V platform, it could feasibly be applied to other architectures (e.g., Arm) as well.

Our advanced PMU provides hardware-based context-aware knowledge of the execution of software layers. We plan to integrate data related to the execution context of applications (e.g., context switches, TLB misses, branches, etc.) with the existing infrastructure that monitors memory- and data-related properties. Our investigations will focus on the balance between software and hardware implementations to get maximal tracing coverage with minimal overheads.

The collected information will serve as input for the detection of attacks that aims at undermining the integrity of AI-based algorithms, as well as for the AI-based NIDS.

As shown in Fig. 1, the PMU-enhanced extension designed to detect adversarial attacks against AI algorithms will operate at the highest privilege level within the software stack. In a software architecture that includes a hypervisor (e.g., [15]), the extension runs upon the hypervisor. Otherwise, the extension will be implemented as a kernel module (e.g., within NuttX or Linux operating systems). In both configurations, the communication between the extension/application and the PMU will be managed via an appropriate driver.

For the development of the extension/application, we will evaluate two distinct approaches. The first approach involves implementing a basic method based on timing behavior and standard statistical analysis. The second approach relies on a K-Nearest Neighbors (KNN) algorithm to determine whether adversarial and benign samples form distinct clusters of PMU data. Both approaches will be evaluated against six different white-box attacks - the same used in the recent empirical evaluation [3]. As timing constraints will not allow us to test this application/extension against the IDS, we will test it against ANNs typically deployed in edge mobile devices. For this purpose, we will borrow, at least, three distinct ANNs from the MLPerf Inference Benchmark [16]. If PMU data is revealed to be affected by adversarial examples, future work might encompass the extension of the framework with an AI algorithm to detect adversarial examples.

Building on the ZeroTrust project, this new framework will explore how to leverage PMU data to enhance the performance of Network Intrusion Detection Systems (NIDS). Our goal is to improve the integrity of drone communication by utilizing OpenTitan (OT). The objective is to develop an efficient AI-enhanced NIDS that operates within a secure enclave. Since OT can perform cryptographic operations, network payloads can be decrypted and analyzed on-the-fly, also correlating this data with PMU information about network stack execution. Specifically, we plan to extend the work of AlSaqr and ZeroTrust by: (i) implementing an AI-based NIDS within OT, focusing on the analysis of attacks in network payloads and extending state-of-the-art approaches [2] using PMU data; and (ii) integrating post-quantum cryptography into OT. The proposed approach offers several advantages: (1) intrusion detection is conducted while preserving communication privacy; (2) the NIDS operates within a secure enclave; and (3) it can leverage PMU information.

References

1. Whelan, J., et al.: Artificial intelligence for intrusion detection systems in unmanned aerial vehicles. Comput. Electr. Eng. (2022). ISSN: 0045-7906. https://doi.org/10.1016/j.compeleceng.2022.107784
2. Li, X., et al.: ROPNN: detection of ROP payloads using deep neural networks. In: CoRR (2018). arXiv: 1807.11110
3. Costa, M., et al.: David and Goliath: an empirical evaluation of attacks and defenses for QNNs at the deep edge. In: 2024 IEEE 9th European Symposium on Security and Privacy (EuroS&P), pp. 524–541 (2024). https://doi.org/10.1109/EuroSP60621.2024.00035.
4. Zoppi, T., et al.: Detect adversarial attacks against deep neural networks with GPU monitoring. IEEE Access (2021). https://doi.org/10.1109/ACCESS.2021.3125920.
5. Ciani, M., et al.: Cyber security aboard micro aerial vehicles: an OpenTitan-based visual communication use case. In: IEEE ISCAS (2023). https://doi.org/10.1109/ISCAS46773.2023.10181732.
6. Parisi, E., et al.: Assessing the performance of OpenTitan as cryptographic accelerator in secure open-hardware system-on-chips. In: ACM CCF (2024). https://doi.org/10.1145/3649153.3649213.
7. Wang, J., et al.: ARI: attestation of real-time mission execution integrity. In: USENIX Security (2023). ISBN: 978-1-939133-37-3
8. Tian, J., et al.: Adversarial attacks and defenses for deep-learning- based unmanned aerial vehicles. IEEE Internet Things J. (2022). https://doi.org/10.1109/JIOT.2021.3111024.
9. Hickling, T., et al.: Robust adversarial attacks detection based on explainable deep reinforcement learning for UAV guidance and planning. IEEE Trans. Intell. Veh. (2023). https://doi.org/10.1109/TIV.2023.3296227.
10. Wang, Z., et al.: A survey on cybersecurity attacks and defenses for unmanned aerial systems. J. Syst. Archit. (2023). ISSN: 1383-7621. https://doi.org/10.1016/j.sysarc.2023.102870. https://www.sciencedirect.com/science/article/pii/S1383762123000498
11. Bouniot, Q., et al.: Optimal transport as a defense against adversarial attacks. In: IEE ICPR (2021). https://doi.org/10.1109/icpr48806.2021.9413327
12. Kadar, M., et al.: Safety-aware integration of hardware-assisted program tracing in mixed-criticality systems for security monitoring. In: IEEE RTAS (2021). https://doi.org/10.1109/RTAS52030.2021.00031.
13. Chen, W., et al.: Low-overhead online assessment of timely progress as a system commodity. In: ECRTS 2023 (2023). ISBN: 978-3-95977-280-8. https://doi.org/10.4230/LIPIcs.ECRTS.2023.13.
14. Zuepke, A., et al.: MemPol: polling-based microsecond-scale percore memory bandwidth regulation. In: Real-Time Systems (2024). https://doi.org/10.1007/s11241-024-09422-8
15. Martins, J., et al.: Bao: a lightweight static partitioning hypervisor for modern multi-core embedded systems. In: Workshop on NG-RES. 2020. ISBN: 978-3-95977-136-8. https://doi.org/10.4230/OASIcs.NG-RES.2020.3.
16. Reddi, V.J., et al.: MLPerf inference benchmark. In: ACM/IEEE ISCA (2020). https://doi.org/10.1109/ISCA45697.2020.00045.

Open Access This chapter is licensed under the terms of the Creative Commons Attribution 4.0 International License (http://creativecommons.org/licenses/by/4.0/), which permits use, sharing, adaptation, distribution and reproduction in any medium or format, as long as you give appropriate credit to the original author(s) and the source, provide a link to the Creative Commons license and indicate if changes were made.

The images or other third party material in this chapter are included in the chapter's Creative Commons license, unless indicated otherwise in a credit line to the material. If material is not included in the chapter's Creative Commons license and your intended use is not permitted by statutory regulation or exceeds the permitted use, you will need to obtain permission directly from the copyright holder.

RL-Enhanced LLMs and Rechargeable Jamming Mines: Achieving Zero-Trust Security for Hierarchical Drone Swarms

Muhammad Shahzad Arif[1(✉)], Sami Muhaidat[2,3], and Paschalis C. Sofotasios[4,5]

[1] Department of Computer and Information Engineering, Khalifa University, Abu Dhabi, UAE
`100062652@ku.ac.ae`
[2] Department of Computer and Information Engineering, KU 6G Center, Khalifa University, Abu Dhabi, UAE
[3] Department of Systems and Computer Engineering, Carleton University, Ottawa, Canada
[4] Department of Computer and Information Engineering, KU C2PS and 6G Center, Khalifa University, Abu Dhabi, UAE
[5] Department of Electrical Engineering, Tampere University, Tampere, Finland

Abstract. This proposal introduces an innovative approach to achieving zero-trust anti-eavesdropping and anti-jamming capabilities in hierarchical drone swarms by employing reinforcement learning (RL) for boosting the potential of telecom-tuned on-device large language models (LLMs). We propose fine-tuning of on-device LLMs based on multi-modal data, including telecom-specific optimization algorithms for autonomous interference/jamming mitigation and anti-eavesdropping measures. To address the lack of emergent behavior in LLMs considering distributed collective intelligence paradigm, we propose a multi-agent RL (MARL) based approach to adaptively optimize policies based on real-time interactions. By treating the swarm of UXVs (i.e. drones) as ultra-dense networks (UDNs) with heterogeneous nodes such as fog or edge drones, we leverage MARL to exploit the inherent interference through optimum node association, resource block selection, and beam/power adjustment at each node. This efficiently mitigates interference at legitimate nodes while enhancing it elsewhere, adhering to a zero-trust approach by assuming eavesdroppers can be located anywhere. We also introduce the concept of rechargeable jamming mines (RJMs) onboard daughter or multi-role (edge) drones, which harvest energy from ambient radio frequency in the UDN environment. These RJMs are deployed and activated at strategic locations to create secure zones, maximizing the secure area around the drones by generating interference that disadvantages eavesdroppers, even with superior channel conditions. Our proposed approach has proven effective in typical UDN scenarios, where the MARL approach is used to optimize configuration settings at each base station, significantly enhancing secure area coverage. Furthermore, the solution is extendable for autonomous anti-jamming capabilities, allowing dynamic channel switching or transmission rate adaptation to mitigate jamming or interference. Our work aligns with the GENZERO24 vision of autonomous secure drone communication, offering robust protection against evolving threats and enhancing operational integrity in complex dynamic environments.

Keywords: Telecom LLMs · reinforcement learning · interference management · anti-eavesdropping · zero trust security

1 Introduction

After the seminal paper on the attention mechanism that completely shifted the natural language processing paradigm from recurrent neural networks to transformer-based architectures [1], there has been a significant surge in the development and application of transformers across various domains. This transition has led to remarkable improvements in language understanding, generation, and translation tasks, paving the way for more sophisticated models like BERT [2], GPT [3], and T5 [4], which leverage the attention mechanism to capture long-term dependencies and contextual information more effectively than their predecessors.

Generative AI (GenAI) is revolutionizing various fields by generating new content based on patterns learned from large datasets. Its integration into wireless networks is anticipated to bring about significant improvements in network design, operation, and management, leading to self-evolving networks that adapt autonomously to changing conditions and demands [5]. The adaptation of large GenAI models into the telecom domain aims to create foundation models trained on multimodal data to perform general telecom-related tasks. By fine-tuning with specific data, these foundation models can handle various tasks such as optimization of modulation, coding, power allocation, and beam-forming [6]. It is important to highlight that adapting GenAI to the edge devices in the telecom domain faces challenges like resource constraints, and model compression which require optimizing models and managing distributed architectures effectively.

On the other hand, reinforcement learning has shown great potential in the real-time optimization of wireless communications. RL is based on learning from interactions with the environment, utilizing reward and penalty mechanisms to improve decision-making in complex scenarios [?]. In the same context, RL is essential for on-device compressed large language models, enabling self-evolving networks that autonomously adapt to environmental changes. A grounding paradigm can be envisioned by combining the strengths of large language models and reinforcement learning into a distributed collective intelligence, resulting in unprecedented security performance [8].

As the focus of GENZERO24 is to revolutionize autonomous systems by integrating GenAI and LLMs within a zero-trust architecture to enhance the security, resilience, and safety of systems managing edge autonomous devices. However, anti-eavesdropping is an area that remains largely unexplored from a zero-trust security perspective. We propose an innovative RL-enhanced zero-trust secure communication framework for hierarchical drone swarms. This framework aims to fine-tune on-device telecom large language models trained on multi-modal data, including RF and telecom-specific data, and employ optimization algorithms for autonomous interference mitigation and anti-eavesdropping. To address the lack of emergent behavior in LLMs, we employ reinforcement learning, enabling LLMs to adaptively optimize policies based on real-time interactions.

2 Proposed Approach

Consider a scenario, as depicted in Fig. 1, where multiple UXVs, i.e., fog and edge drones, are deployed in a hierarchical swarm architecture. We also consider RJMs onboard daughter drones, which harvest energy from ambient RF in the dense swarm environment and can be activated to create interference at appropriate locations where potential eavesdroppers may be present. It is important to highlight that for approaching the problem from a zero-trust architecture perspective, we focus on mitigating interference at legitimate nodes while enhancing interference elsewhere by considering that eavesdroppers can be located anywhere. We consider that each fog or edge drone acts as a distributed agent, controlling multiple daughter drones (with onboard RJMs) under its command to effectively manage their deployment, charging, and activation on an as-needed basis. Each agent has a compressed telecom-tuned on-device LLM installed.

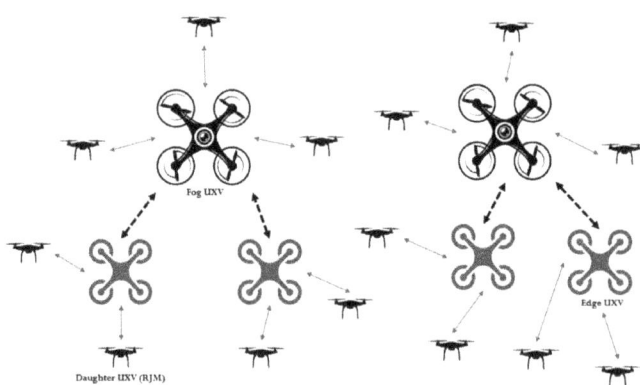

Fig. 1. Hierarchical drone swarm architecture with RJM based daughter drones.

We propose an RL-based capability enhancement of fine-tuned on-device telecom-LLMs to create a zero-trust secure hierarchical drone architecture. Firstly, we aim to fine-tune the on-device LLMs using multi-modal data, which includes RF data, textual data, and optimization algorithms focused on security, interference mitigation, and anti-eavesdropping capability. This training, however, lacks emergent behavior. To address this, we employ RL to enable the LLMs to learn from real-time interactions, adapt decision-making policies, and grant the system autonomous zero-touch self-healing and self-evolving capabilities.

By modeling the swarm of UXVs (drones) as ultra-dense networks with diverse node types, including fog and edge drones, we utilize MARL to handle inherent interference resulting from resource block reuse. This is achieved through intelligent node association, resource block allocation, and beam/power adjustments for each node. This approach effectively reduces interference at legitimate nodes while intentionally increasing it in areas where eavesdroppers might be present, following a zero-trust security model that assumes threats could be located anywhere. The rechargeable jamming mines can be installed onboard daughter or multi-role drones (fog drones) to harvest energy from ambient RF in the UDN environment. These can be strategically deployed to strengthen secure zones, expanding the secure area around the drones and effectively countering eavesdroppers, even those with advantageous channel conditions.

3 Preliminary Results

We have demonstrated the effectiveness of our proposed approach in a typical UDN scenario. As network densification leads to aggressive frequency reuse, inter-cell interference rises, which can be exploited to enhance interference in locations likely to have eavesdroppers. By intelligently managing user association, resource block selection, and the charging/activation of RJMs, inherent interference can be effectively exploited to our advantage. Our approach does not rely on the eavesdropper's location or channel state information and differs from traditional maximum signal to interference and noise ratio (SINR) based user association, focusing instead on maximizing the secure area [?], [10]. We employ MARL-based optimization to maximize secure areas in UDNs through intelligent user association, interference coordination, and resource allocation. Each base station is considered an agent with multiple RJMs in its vicinity that are under its control. We defined the metric of secure area percentage (SAP) which drives the reward shaping in our RL-based optimization framework. Our goal is to maximize the secure area while minimizing resource expenditure and maintaining the desired minimum quality of service (QoS) requirements for all legitimate users. The judicious allocation of resources is maintained through reward shaping, achieved by integrating penalties into the reward definition. This ensures that the converged solution, specifically the configuration settings of base stations, prioritizes both energy and resource efficiency.

Our approach has been validated through extensive simulations using a multi-agent deep RL architecture that utilizes only local state information. We achieved an average SAP of up to 90% while ensuring that legitimate user QoS requirements are met in the considered scenario. This indicates that the proposed solution is suitable for implementation on edge devices due to the comparatively much smaller neural network architecture compared to a single (central) agent.

The results are depicted in Fig. 2, where we simulated three agents i.e. base stations, each controlling two jamming mines. The scenario had four user nodes. The DRL agent (base station) trains independently, however, a common reward is shared with all agents to boost collaboration. Every action is attributed to a specific configuration setting of the BS i.e. the resource block, antenna orientation, beam type, and output power to effectively manage the interference and charging/activation of the associated RJMs to maximize the overall secure area while ensuring appropriate user association with required QoS at all legitimate users. We employed $\varepsilon-$greedy policy to effectively learn the optimized sequence of actions that achieves long-term higher secure area percentage [11]. As shown in Fig. 2a, after convergence (i.e., during the exploitation phase), our multi-agent RL-based optimization approach secures up to 90% of the area. It is important to note that the optimal solution also ensures that all legitimate users meet the required QoS standards. As shown in Fig. 2b, an average capacity reward of 100 indicates that the QoS requirements for all users are satisfied.

It is worth noting that the achievable SAP is constrained by the relative locations of base stations, user nodes, and jamming mines, which were considered fixed in our simulation settings. However, in drone swarm use cases, this limitation would be mitigated by strategically deploying daughter drones, which improves the secure area percentage and thus supports the realization of the envisioned zero-trust architecture.

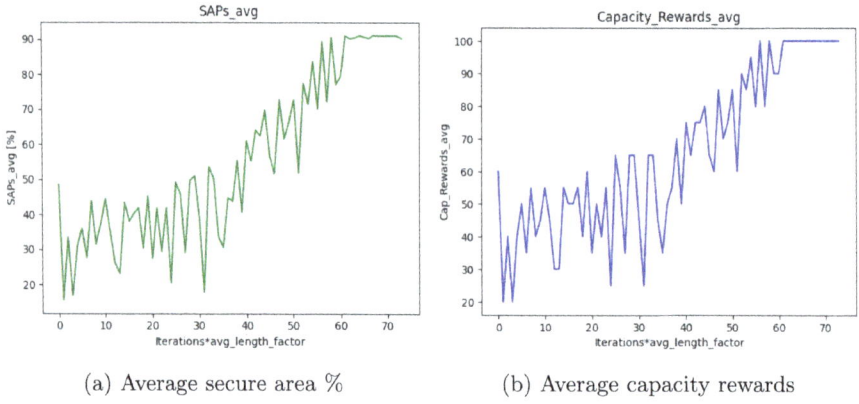

(a) Average secure area %   (b) Average capacity rewards

Fig. 2. Average secure area % and capacity rewards improvement with multi-agent RL approach.

Our future work will adapt the current framework to dynamic environments like drone swarms, focusing on advancing autonomous, secure communication and developing a zero-trust security model in complex, evolving scenarios. Leveraging GenAI and RL, we will extend the existing work to include autonomous anti-jamming capabilities, enabling dynamic channel switching and transmission rate adaptation in response to interference. Through multi-agent reinforcement learning, the system will optimize these adaptations in real-time, ensuring robust communication and enhanced network resilience even under jamming attacks. Moreover, we aim to quantify the limits of the complexity involved in various scenarios of interest, considering realistic technical and operational conditions. This thorough and comprehensive analysis is expected to highlight the associated limitations of the considered scenarios and the field as a whole, providing valuable insights for both the theoretical design and practical deployment of such systems.

4 Conclusion

In conclusion, this proposal introduces an innovative zero-trust communication framework for hierarchical drone swarms, integrating fine-tuned on-device telecom LLMs with reinforcement learning to optimize real-time interactions and autonomous decision-making. By leveraging multi-agent RL for intelligent node association and interference management, combined with rechargeable jamming mines, we aim to achieve zero-trust, anti-eavesdropping capabilities. Our preliminary framework, applied to a typical UDN scenario, has demonstrated a significant increase in secure area coverage. Future work will extend this framework to dynamic drone swarms, advancing secure communication and zero-trust security with the aid of GenAI.

References

1. Vaswani, A., et al.: Attention is all you need. Adv. Neural Inf. Process. Syst. **30** (2017)
2. Devlin, J., Chang, M.-W., Lee, K., Toutanova, K.: BERT: pre-training of deep bidirectional transformers for language understanding. arXiv preprint arXiv:1810.04805 (2019)
3. Brown, T.B., et al.: Language models are few-shot learners. arXiv preprint arXiv:2005.14165 (2020)
4. Raffel, C., et al.: Exploring the limits of transfer learning with a unified text-to-text transformer. J. Mach. Learn. Res. **21**(140), 1–67 (2020). https://arxiv.org/abs/1910.10683
5. Chafii, M., et al.: Twelve scientific challenges for 6g: rethinking the foundations of communications theory. IEEE Commun. Surv. Tutor. (2023)
6. Bariah, L., Zhao, Q., Zou, H., Tian, Y., Bader, F., Debbah, M.: Large generative AI models for telecom: the next big thing? arXiv preprint arXiv:2306.10249 (2023). https://arxiv.org/abs/2306.10249
7. Luo, F.-L.: Machine Learning for Future Wireless Communications. Wiley-IEEE Press (2020)
8. Carta, T., Romac, C., Wolf, T., Lamprier, S., Sigaud, O., Oudeyer, P.-Y.: Grounding large language models in interactive environments with online reinforcement learning. arXiv preprint arXiv:2302.02662 (2023)
9. Dania Marabissi, S.C., Mucchi, L.: Physical-layer security metric for user association in ultra-dense networks. In: 2020 Int. Conf. on Computing, Networking and Communications (ICNC), pp. 487–491 (2020). https://api.semanticscholar.org/CorpusID:214761845

10. Marabissi, D., Abrardo, A., Mucchi, L.: A new framework for physical layer security in HetNets based on radio resource allocation and reinforcement learning. Mobile Netw. Appl., 1–9 (2023)
11. Sutton, R.S., Barto, A.G.: Reinforcement Learning: An Introduction. MIT Press (2018)

Open Access This chapter is licensed under the terms of the Creative Commons Attribution 4.0 International License (http://creativecommons.org/licenses/by/4.0/), which permits use, sharing, adaptation, distribution and reproduction in any medium or format, as long as you give appropriate credit to the original author(s) and the source, provide a link to the Creative Commons license and indicate if changes were made.

The images or other third party material in this chapter are included in the chapter's Creative Commons license, unless indicated otherwise in a credit line to the material. If material is not included in the chapter's Creative Commons license and your intended use is not permitted by statutory regulation or exceeds the permitted use, you will need to obtain permission directly from the copyright holder.

Challenge 5: – Human-Swarm Collaboration Interface for Enhanced Drone and Swarm Resilience

Symbolic Constraint-Solving Capabilities of Transformer Large Language Models

Leyan Pan[1(✉)], Chris Esposo[2], Jacob Abernethy[2], Vijay Ganesh[2], and Wenke Lee[1]

[1] School of Cybersecurity and Privacy, Georgia Institute of Technology, Atlanta, Georgia
{leyan.pan,wenke.lee}@gatech.edu
[2] School of Computer Science, Georgia Institute of Technology, Atlanta, Georgia
{cesposo,jabernethy,vganesh}@gatech.edu

Abstract. The empirical success of Transformer-based Large Language Models (LLMs) has elicited significant interest in their application for security domains such as formal verification and program analysis. However, their application in formal domains, such as formal verification and program analysis, is constrained by the lack of mechanical understanding and the risk of hallucinations. To address these challenges, we present a robust theoretical construction establishing that transformer-LLMs with CoT can solve an NP-Hard logical reasoning problem, specifically 3-SAT. Our construction provides theoretical guarantees that these models can either simulate the deductive search processes of SAT-solving or probabilistically generate correct solutions as the most likely completion. Empirically, we demonstrate that transformers can generalize these procedures across different formula distributions, reinforcing their potential for reliable formal reasoning.

Keywords: Symbolic constraint solving · Large Language Models · Transformers · SAT

1 Introduction

Transformer-based Language Models (LLMs) [4] have demonstrated remarkable success in natural language reasoning tasks, especially when using prompting techniques such as Chain-of-Thought (CoT, [5]), and have been proposed for various security applications. However, recent research reveals that even the most advanced LLMs face challenges in reliable reasoning, often producing hallucinations and struggling with length generalization. These limitations are particularly concerning in security-critical applications such as formal verification of autonomous systems with strict requirements on accuracy. To fundamentally understand Transformer's reliability in security applications, we rigorously investigate the SAT-solving capabilities of Transformer Large Language Models (LLMs), focusing specifically on SAT-solving.

An SAT solver determines the 'satisfiability' of a boolean set of equations for a set of inputs, while an SMT solver (Satisfiability Modulo a Theory) extends this capability to more complex data structures such as bit-vectors, strings, arrays, and beyond. These tools are essential for reducing a program to its logical components, allowing us to prove its satisfiability or identify concrete counterexamples. Through this process, we can analyze program properties and uncover unreachable paths, error conditions, and security vulnerabilities, making them crucial for enhancing the security and reliability of software systems.

Contributions. We prove by theoretical construction that decoder-only Transformers can solve 3-SAT, a fundamental NP-Complete logical reasoning problem, by performing logical deduction and backtracking using Chain-of-Thought (CoT). We show that Transformers can perform logical deduction on all conditions (clauses) in parallel instead of checking each condition sequentially. Nevertheless, the construction requires exponentially many CoT steps in the worst case, although it is much faster on most typical examples. Our training experiments suggest that Chain-of-Thought allows Transformer-LLMs to achieve out-of-distribution generalization for the same input lengths.

Significance. Our results advance our theoretical understanding of the capabilities of LLMs in performing reliable NP-Hard logical reasoning problems critical to formal verification and program analysis. Our observations on the capabilities and limitations of Transformer-LLMs can guide the development of compound LLM-integrated Cyber-reasoning systems that leverage the abstract reasoning capabilities of LLMs to perform high-level analysis of security threats (Fig. 1).

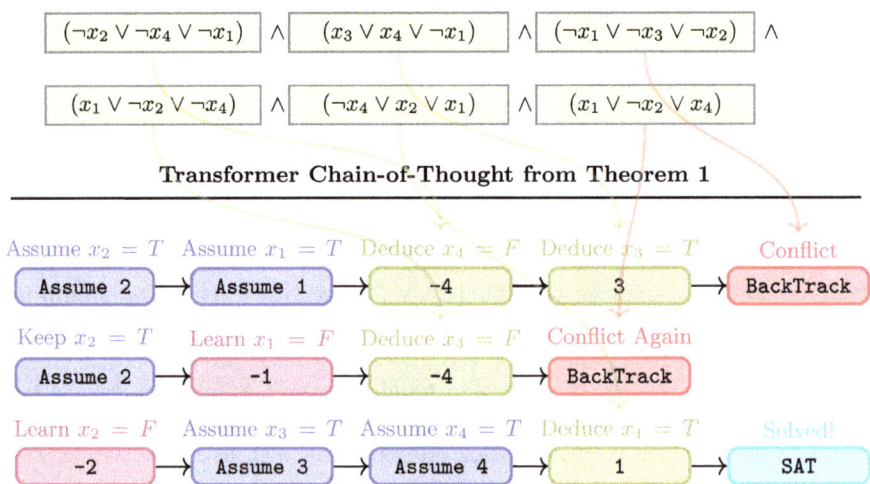

Fig. 1. Visualization of the Chain-of-Thought (CoT) process used by our model to solve a sample SAT formula. The model autonomously performs trial-and-error reasoning, backtracking upon conflicts.

2 Preliminaries

2.1 Problem Formulation

Our paper focuses on the capabilities of autoregressive decoder-only Transformers in deciding the satisfiability of 3-SAT instances using Chain-of-Thought (CoT).

3-SAT. An SAT (Boolean Satisfiability) instance is a formula of boolean variables connected with logical operators AND, OR, and NOT. 3-SAT is a specific form of the SAT problem where the Boolean formula is expressed in conjunctive normal form (CNF) with three literals per clause. A formula in CNF is a conjunction (i.e. "AND") of clauses, a **clause** is a disjunction (i.e. "OR") of several **literals**, and each literal is either a variable or its negation. In the case of 3-SAT, each clause contains exactly three literals. Formally, a 3-CNF formula ϕ can be written as:

$$\phi = (l_{1,1} \vee l_{1,2} \vee l_{1,3}) \wedge (l_{2,1} \vee l_{2,2} \vee l_{2,3}) \wedge \cdots \wedge (l_{c,1} \vee l_{c,2} \vee l_{c,3})$$

where $l_{i,j}$ is a literal, $\wedge$ represents AND, $\vee$ represents OR, and c is the number of clauses. An example 3-SAT formula with 4 variables and 6 clauses is:

$$(x_1 \vee \neg x_2 \vee x_3) \wedge (\neg x_1 \vee x_2 \vee \neg x_3) \wedge (x_2 \vee x_4 \vee \neg x_1) \wedge$$
$$(x_1 \vee \neg x_3 \vee x_4) \wedge (\neg x_2 \vee \neg x_3 \vee \neg x_4) \wedge (x_3 \vee \neg x_4 \vee \neg x_1)$$

The 3-SAT problem refers to determining if any assignment of truth values to the variables allows the formula ϕ to evaluate as true. It is well-known that both SAT and 3-SAT are NP-hard and are widely conjectured to be unsolvable in polynomial time.

2.2 Autoregressive Decoder-only Transformer Model

A decoder-only Transformer takes a sequence of input tokens as a prompt and repeatedly predicts the next token to generate a response. For theoretical results, we consider Transformers with learned positional encodings, ReGLU as MLP activation, and ignore the normalization operations. Specifically, given an input sequence s of length n, the Transformer operates as follows:

Token Embedding and Positional Encoding. Each input token s_i ($i \in [n]$) is converted into a d-dimensional vector $e_i = \text{Embed}(s_i) \in \mathbb{R}^d$ using an embedding layer. To retain sequence order, positional embeddings $p_i \in \mathbb{R}^d$ are added, resulting in $X^{(0)} = [e_1 + p_1, \cdots, e_n + p_n]^\top \in \mathbb{R}^{n \times d}$.

Transformer Blocks. The sequence is processed through L Transformer blocks, each transforming the input as follows:

$$H^{(l)} = X^{(l-1)} + \text{MHA}^{(l)}(X^{(l-1)})$$
$$X^{(l)} = H^{(l)} + \text{FFN}^{(l)}(H^{(l)})$$

where $\text{MHA}^{(l)}$ and $\text{FFN}^{(l)}$ denote the multi-head self-attention layer and feed-forward network, respectively:

$$\text{MHA}^{(l)}(X) = \sum_{h=1}^{H} A^{(l,h)} X W_V^{(l,h)} W_O^{(l,h)}, \tag{1}$$

$$A^{(l,h)} = \text{softmax}\left(X W_Q^{(l,h)} (X W_K^{(l,h)})^\top + M\right) \tag{2}$$

$$\text{FFN}^{(l)}(X) = \sigma(X W_1^{(l)}) W_2^{(l)}. \tag{3}$$

Here, $W_Q^{(l,h)}, W_K^{(l,h)}, W_V^{(l,h)}, W_O^{(l,h)}$ are the query, key, value, and output matrices of the h-th head and l-th layer, and $W_1^{(l)}, W_2^{(l)}$ are the FFN weight matrices. The activation function σ is ReGLU [2] for our theoretical construction, where $x \in \mathbb{R}^{2d}$ is split into the concatenation of $x_1, x_2 \in \mathbb{R}^{2d}$ and

$$\sigma(x) = x_1 \otimes \text{ReLU}(x_2)$$

where $x = x_1 \circ x_2$ and $\otimes$ denotes element-wise product. The causal mask $M \in \{-\infty, 0\}^{n \times n}$ ensures each position i only attends to preceding positions $j \leq i$, which is essential for autoregressive generation.

Autoregressive Decoding and Chain-of-Thought. During generation, the Transformer model is repeatedly invoked to generate the next token and appended to the input tokens. When using the greedy seconding strategy, the highest probability token is always selected and the generation process is described in Algorithm 1.

Algorithm 1: Greedy Decoding

Input: Model $M : \mathcal{V}^* \to \Delta(\mathcal{V})$, prompt $s_{1:n} = (s_1, s_2, \ldots, s_n)$, stop tokens $\mathcal{E} \subseteq \mathcal{V}, t \leftarrow n$
1 **while** $t \leftarrow t+1$ **do**
2     $\mathbf{p}_t \leftarrow M(s_{1:t-1})$;   // Obtain model output
3     $s_t \leftarrow \arg\max_{v \in \mathcal{V}} \mathbf{p}_t(v)$;   // Select most probable token
4   **if** $s_t \in \mathcal{E}$ **return** $s_{1:t}$
5 **end**

Table 1. Average accuracies (%) of SAT/UNSAT prediction for models trained and tested on different datasets in two training regimes: $p \in [6, 10]$ and $p \in [11, 15]$. Each accuracy is computed over 10000 samples.

	$p \in [6, 10]$			$p \in [11, 15]$		
	Marginal	Random	Skewed	Marginal	Random	Skewed
Marginal	99.88%	99.99%	99.99%	98.66%	99.70%	99.57%
Random	99.96%	100.00%	100.00%	99.11%	99.75%	99.55%
Skewed	99.96%	100.00%	99.99%	99.41%	99.74%	99.48%

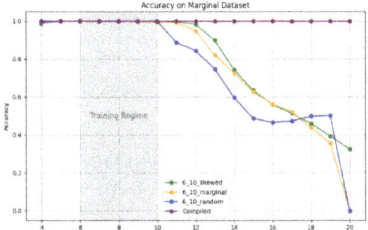 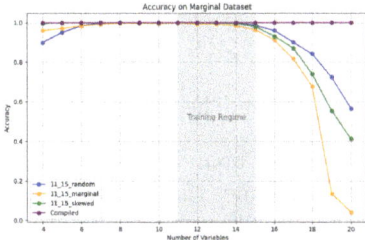

Fig. 2. Result of the length-generalization experiments, showing SAT/UNSAT prediction accuracy of Transformer-based LLMs trained on different dataset distributions. (Left) model trained on $6 \leq p \leq 10$, (Right) model trained on $11 \leq p \leq 15$.

3 Expressiveness of Auto-Regressive Transformers in Logical Constraint Solving

This section presents and explains our main results on Transformers' capability in deductive reasoning and backtracking with CoT. To rigorously state our results, we first formally define decision problems, decision procedures, and what it means for a model to "solve" a decision problem using CoT:

Definition 1 (Decision Problem). *Let $\mathcal{V}$ be a vocabulary, $\Sigma \subseteq \mathcal{V}$ be an alphabet, $L \subseteq \Sigma^*$ be a set of valid input strings. We say that a mapping $f : L \to \{0, 1\}$ is a decision problem defined on L.*

Definition 2 (Decision Procedure). *We say that an algorithm $\mathcal{A}$ is a decision procedure for the decision problem f, if given any input string x from L, $\mathcal{A}$ outputs 1 if $f(x) = 1$, and 0 otherwise.*

Definition 3 (Autoregressive Decision Procedure). *For any map $M : \mathcal{V}^* \to \Delta(\mathcal{V})$, which we refer to as an auto-regressive next-token prediction model, and $\mathcal{E} = \{\mathcal{E}_0, \mathcal{E}_1\} \subset \mathcal{V}$, define decision procedure $\mathcal{A}_{M,\mathcal{E}}$ as follows: For any input $s_{1:n}$, run Algorithm 1 with stop tokens $\mathcal{E}$. $\mathcal{A}_{M,\mathcal{E}}$ outputs 0 if $s_{1:t}$ ends with $\mathcal{E}_0$ and $\mathcal{A}_{M,\mathcal{E}}$ output 1 otherwise. We say M autoregressively decides decision problem f if there is some $\mathcal{E} \subset \mathcal{V}$ for which $\mathcal{A}_{M,\mathcal{E}}$ decides f.*

Definition 4 (3-SAT$_{p,c}$). *Let DIMACS(p, c) denote the set of valid DIMACS encodings of 3-SAT instances with at most p variables and c clauses with a prepended [BOS] token and an appended [SEP] token. Define 3-SAT$_{p,c}$: DIMACS$(p, c) \to \{0, 1\}$ as the problem of deciding whether the 3-SAT formula encoded in the input in DIMACS(p, c) encoding is satisfiable.*

With the above definition, we're ready to present a formal statement of our theoretical construction of a Transformer model that performs SAT Solving:

Theorem 1 (Decoder-only Transformers can solve SAT). *For any $p, c \in \mathbb{N}^+$, there exists a Transformer model $M : \mathcal{V}^* \to \Delta(\mathcal{V})$ that autoregressively decides 3-SAT$_{p,c}$ in no more than $p \cdot 2^{p+1}$ CoT iterations. M requires $L = 7$ layers, $H = 5$ heads, $d_{emb} = O(p)$, and $O(p^2)$ parameters.*

Remarks on Theorem 1

- The upper bound on the CoT length $p \cdot 2^{p+1}$ is a worst-case upper bound which assumes that the model is unable to make any logical deductions have to try all 2^p assignments. However, this upper bound is never reached in practice, and number of CoT tokens is no greater than $8p \cdot 2^{0.08p}$ for most random formulas. If the number of backtracking steps is bounded by T then the CoT is no longer than $(2p + 1)(T + 1)$
- The worst-case CoT length is independent of the number of clauses c, which is due to the parallel deduction over all clauses within the Transformer construction. Otherwise, sequentially processing each clause would take at least $c \cdot 2^{O(p)}$ number of steps.
- Positional encodings are not included in the number of parameters. The positional encoding at position i is the numerical value i at a particular dimension.
- Each parameter can be represented with $O(p + \log c)$ bits

The construction uses adapted versions of lemmas from [1] as basic building blocks.

Here we describe our design of the algorithm and CoT process for SAT-Solving and a high-level description on the internal mechanisms of the Transformer model.

4 Empirical Results

Based on our theoretical construction, we conduct experiments to show whether LLMs can learn SAT-Solving with our theoretical CoT design through training.

Datasets and Models. We construct three different types of training datasets, each including 50,000 samples, comprising three distributions of CNF-SAT formulas for training on 6–10 variables. These datasets, listed in order of difficulty, are:

- **Marginal 3-SAT (Hardest):** Composed of pairs of formulas that differ by only one token. The main purpose is to reduce the effectiveness of leveraging statistical features in the formulas.
- **Random:** Formulas not paired by differing tokens and each clause is randomly generated
- **Skewed:** Formulas where polarity and variable sampling are not uniform; one polarity is preferred over the other with probability 0.75, and some literals are preferred over others.

From the experiments, we a miniaturized LLaMa [3] model for RoPE-Encoding.

4.1 Intra-Length OOD Generalization

Our first set of experiments evaluates the model's performance on SAT formulas sampled from different distributions from training, but the number of variables in formulas remains the same ($p \in [6, 10]$ and $p \in [11, 15]$ for both train and test datasets).

As shown in Table 1, our trained models achieve near-perfect SAT vs UNSAT prediction accuracy when tested on the same number of variables as the training data, even when on formulas sampled from different distributions. Recall that the "marginal" dataset has SAT vs UNSAT samples differing by a single token (out of at least $16p$ tokens in the input formula), which minimizes statistical evidence that can be used for SAT/UNSAT prediction. Our experiments suggest that the LLM have very likely learned general reasoning procedures using CoT that can be applied to all formulas with the same number of variables as the data they are trained on.

4.2 Length Generalization

The second experiment evaluates the model's ability to generalize to formulas with a different number of variables than seen during training. We use the model trained on the Marginal dataset from Sect. 4.1 and evaluate datasets with 4–20 variables, generated using the three methods described, with 2,000 samples each. For this experiment, we evaluate the binary SAT vs UNSAT prediction accuracy.

Results. In Fig. 2, our results indicate that performance degrades drastically beyond the training regime when the number of variables increases. This shows that the model is unable to learn a general SAT-solving algorithm that works for all inputs of arbitrary lengths, which corroborates our theoretical result where the size of the Transformer for SAT-solving depends on the number of variables. This further demonstrates the value of having a compiled Transformer that provably works well on all inputs up to p variables for any given p.

References

1. Feng, G., Zhang, B., Gu, Y., Ye, H., He, D., Wang, L.: Towards revealing the mystery behind chain of thought: a theoretical perspective. In: Oh, A., Naumann, T., Globerson, A., Saenko, K., Hardt, M., Levine, S. (eds.) Advances in Neural Information Processing Systems 36: Annual Conference on Neural Information Processing Systems 2023, NeurIPS 2023, New Orleans, LA, USA, 10–16 December 2023 (2023)
2. Shazeer, N.: GLU variants improve transformer. CoRR **abs/2002.05202** (2020), https://arxiv.org/abs/2002.05202
3. Touvron, H., et al.: Llama: open and efficient foundation language models. ArXiv **abs/2302.13971** (2023)
4. Vaswani, A., et al.: Attention is all you need. In: Guyon, I., et al. (eds.) Advances in Neural Information Processing Systems 30: Annual Conference on Neural Information Processing Systems 2017, 4–9 December 2017, Long Beach, CA, USA, pp. 5998–6008 (2017)
5. Wei, J., et al.: Chain-of-thought prompting elicits reasoning in large language models. In: Koyejo, S., Mohamed, S., Agarwal, A., Belgrave, D., Cho, K., Oh, A. (eds.) Advances in Neural Information Processing Systems, vol. 35, pp. 24824–24837. Curran Associates, Inc. (2022)

Open Access This chapter is licensed under the terms of the Creative Commons Attribution 4.0 International License (http://creativecommons.org/licenses/by/4.0/), which permits use, sharing, adaptation, distribution and reproduction in any medium or format, as long as you give appropriate credit to the original author(s) and the source, provide a link to the Creative Commons license and indicate if changes were made.

The images or other third party material in this chapter are included in the chapter's Creative Commons license, unless indicated otherwise in a credit line to the material. If material is not included in the chapter's Creative Commons license and your intended use is not permitted by statutory regulation or exceeds the permitted use, you will need to obtain permission directly from the copyright holder.

Cognitive-Aware Multi-modal Human-Swarm Interface for Optimal Collaboration

Sooyung Byeon(✉), Joonwon Choi, and Inseok Hwang

Purdue University, West Lafayette, IN 47907, USA
{sbyeon,choi774,ihwang}@purdue.edu

Abstract. This paper presents the development of an interface that enables humans to collaborate with robot swarms (RSs) to effectively accomplish missions in highly uncertain environments. Autonomous RS technology has advanced to the point where robots can perform various tasks without human intervention. However, there are still two major uncertainties in mission execution. The first uncertainty arises from the limitations of artificial intelligence (AI), such as the inaccuracy of computer vision systems. The second uncertainty comes from the decision-making aspect, where the mission requirements may not be accurately communicated to the RS. To overcome these uncertainties, it is crucial to design a system that allows humans to monitor the RS and actively participate in critical decision-making processes during missions. We design an interface to: 1) estimate human cognitive states to monitor the human partner, 2) determine the best communication methods to calibrate cognitive states, and 3) allocate functions among humans and robots to optimize team performance. We demonstrate this approach through drone search and rescue (SAR) missions.

Keywords: human-AI collaboration · robot swarm · cognitive state · multi-modal interface · function allocation

1 Introduction

Human-AI collaboration (HAC) combines the strengths of human cognitive capabilities and the accurate computational abilities of artificial intelligence (AI) to efficiently perform complex missions in dynamic and highly uncertain environments [1]. For example, in the manufacturing sector, particularly with Industry 4.0, research has focused on ways for robots and humans to share the same physical space and divide sub-tasks among themselves to accomplish a larger task [2]. In the field of vehicle control, shared control methods where humans and AI work together to perform complex maneuvers have been widely studied [3]. Additionally, human-in-the-loop research has been actively conducted to ensure that humans can participate in decision-making in a timely manner.

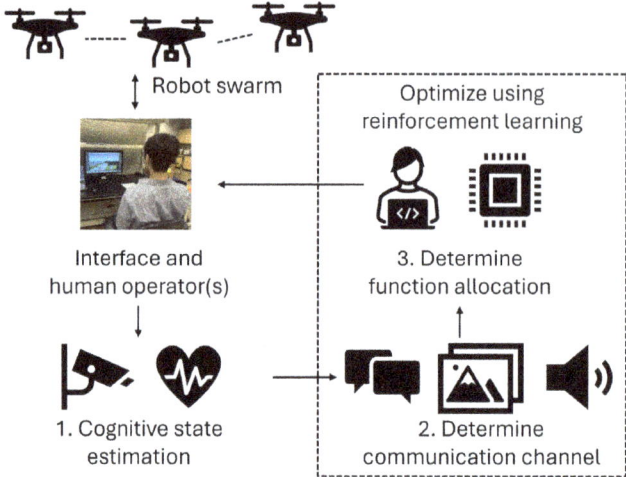

Fig. 1. The proposed human-swarm interface architecture consists of cognitive state estimation, communication channels, and function allocation.

Many studies have also focused on maintaining human cognitive states in a way that enhances close collaboration with AI [4].

For genuine collaboration, effective communication between all entities is crucial. Traditional HAC research has primarily focused on one-on-one or small group communication between humans and robots. However, with the emerging challenges of managing robot swarms (RSs) such as unmanned vehicle fleets and urban air mobility (UAM), where a few humans oversee many robots, there is a growing need to formally address the scalability of team communication. For instance, during a disaster where a RS performs search and rescue (SAR) missions, it is difficult for a human to monitor every robot. Robots must independently distribute among themselves and perform tasks, selectively relaying only essential information to humans. Humans need to set overall mission goals, understand the changing situation, make timely decisions, and issue necessary commands accordingly. Moreover, many RS operations, including SAR missions, take place in complex, uncertain, and unstructured environments. Therefore, determining how humans can efficiently collaborate with multiple robots to enhance team performance and safety is a critical issue.

2 Problem Definition

Our goal is to develop a human-swarm interface architecture that monitors human cognitive states, optimizes multi-modal communication channels, and allocate functions to ensure efficient and safe human-swarm collaboration as shown in Fig. 1. The problem can be decomposed into three sub-problems.

The first sub-problem involves collecting, analyzing, and learning from human behavioral and physiological data to numerically estimate the human cognitive state. Understanding the counterpart's condition is a crucial starting point for effective human-swarm team communication. If AI is unaware of the human cognitive state, effective collaboration among them is difficult to achieve [4]. Therefore, our proposed architecture monitors humans and responds interactively. To avoid restricting human behavior during collaboration, we use non-invasive sensors such as face recognition camera and heart rate sensor (Sect. 3).

The second sub-problem is to establish multi-modal communication channels that allow fast interaction between humans and robots. Advances in the large language model (LLM) technology enable robots to engage in verbal communication with humans. While maintaining the concise information delivery capabilities of auditory and visual interfaces widely used in existing HAC research, we propose to utilize the latest LLM technology to facilitate the exchange of commands and feedback between humans and robots (Sect. 4).

The final sub-problem is to optimize function allocation between humans and robots, i.e., who does what and when [5]. Function allocation can leverage human's knowledge based behaviors and robot's skill and rule based behaviors. Based on the current human cognitive state and mutual understanding through communication, we can derive an optimized function allocation policy. This function allocation must be adaptive to both humans and situations to optimize overall team performance and respond flexibly to circumstances (Sect. 5).

3 Cognitive State Estimation

People know from experience that effective communication is difficult to achieve when the interaction counterpart is busy with other tasks (i.e., overloaded). Additionally, following the decisions of someone who does not have an accurate understanding of the situation can jeopardize the team's safety. Therefore, we aim to develop methods for estimating the workload in the proposed architecture. We further plan to extend our solid foundation in cognitive state modeling approaches to address other cognitive states such as situational awareness and trust [6].

3.1 Workload

Workload represents the cognitive load a person feels while performing a task. We utilize the latest cognitive architectures to develop components that estimate workload in real-time based on the situation. There have been several works to estimate human workload based on physiological sensors. For instance, electrocardiogram (ECG) and infrared image were utilized along with features extracted using an open source software [7,8]. Motivated by this, we train models using deep learning techniques to understand how workload changes when a human collaborates with multiple robots simultaneously, communicates in real-time, and performs various functions intermittently. Our goal is to maintain the workload of humans at an optimal level, not too high and not too low (i.e., workload calibration), to ensure that humans can interact with robots in the best possible state.

3.2 Situational Awareness and Trust

As future work, we plan to extend our proposed architecture to encompass situational awareness and trust. The widely accepted theory of situational awareness categorizes human understanding of the current situation into three levels [9]. Level 1 is when a human perceives the situation, Level 2 is when a human comprehends the situation, and Level 3 is when a human can project future status based on this understanding. Quantitative models of situational awareness have been extensively studied based on this theory. Using the latest research, we plan to analyze the relationships between situational awareness, other cognitive states, and team performance. Based on this analysis, we determine the methods and depth of communication and function allocation.

The relationship between the transparency of communication (between humans and robots) and trust has been widely studied [10]. For instance, if a robot (AI) provides explanations to humans based on high transparency regarding the current situation and future predictions, it can help humans achieve Level 3 situational awareness, leading to increased trust in AI. We are planning to train and analyze a deep learning model to understand the correlations between various features such as human facial expressions, heart rate, and eye movements, and their trust levels.

3.3 Trade-Off Between Cognitive States and Performance

We note that simply increasing transparency and trust cannot be the only approach for effective mission performance between humans and RS. This is because communication transparency, workload, and situational awareness have complex trade-off relationships. If transparency is increased for all aspects of RS operations, it may lead to human overload, ultimately undermining humans' situational awareness and trust. Such situations can negatively impact function allocation and limit the human's ability to perform effectively as a team member. Therefore, we utilize computational cognitive models which we have developed [11] to numerically analyze these trade-offs and aim to validate and improve them through human user studies.

4 Communication Channels

State-of-the-art LLM technology enables humans to communicate with RSs using natural language which is one of the most intuitive methods. When human commands are abstract or implicit, inferring human intention can be challenging, but translating human commands into mathematically formal methods like temporal logic can help prevent misunderstandings between humans and RS [12]. Hence, we plan to represent the state and control objectives of the RS using generate natural language to effectively communicate this to humans.

However, natural language is not the only effective communication method. For instance, in cases where a robot's path is unsafe, displaying relevant information on a graphical user interface can be a faster and more efficient communication method compared to complex natural language feedback provided by AI.

Hence, we will maintain the use of auditory and visual communication as supplementary tools while researching how communication in complex environments can be conducted through natural language.

5 Function Allocation

Function allocation involves making decisions about which functions each team member should perform to effectively achieve the team's goals [5]. Our research group has studied adaptive function allocation techniques based on a computational model which is formulated as the (partially observable) Markov decision process to optimize team performance and cognitive states [6]. For function allocation to be effective, decisions must consider the relationship between human cognitive states and transparency. While these complex relationships are challenging to model mathematically, we have built a reliable simulation platform using the computational cognition-work model (CCWM) [11]. We can determine the components of the human-swarm interface based on the CCWM model. Subsequent human user studies would allow us to precisely estimate the model parameters and verify the accuracy of the model. This model is designed to be extensible, not just for specific missions. The proposed architecture then can optimize human-swarm collaboration using a reinforcement learning technique by determining two controllable variables: the communication channel and function allocation policy as shown in Fig. 1.

6 Hardware Implementation

To validate the proposed architecture in an environment similar to real-world operations, we have developed interface software integrated with a drone swarm hardware system. For rapid development and iterative testing, we are operating multiple Crazyflie drones in an indoor environment and create a mixed reality environment to simulate complex drone operations as shown in Fig. 2. Using a motion capture system (Qualisys), we can obtain millimeter-level precision in real-time drone position data. Additionally, we have secured software development kits that allow real-time communication with various physiological sensors (e.g., Polar H-10 heart rate sensor and webcam) to create integrated software that combines graphical user interfaces, sensor data, and drone data. The hardware implementation presented in this paper is an extension of our previous work [13].

7 Preliminary Results

We conducted a pilot study to demonstrate the proposed approach, using a drone SAR scenario where a human collaborates with two drones to locate victims in a mountain area. The drones, equipped with cameras, are tasked with visiting six target points and autonomously planning their trajectories using the rapidly

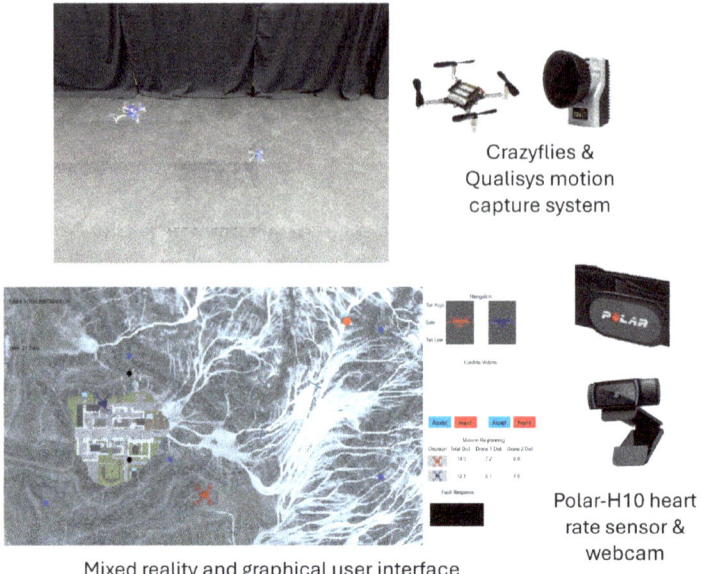

Fig. 2. The hardware elements for developing the proposed human-swarm collaboration interface.

exploring random tree (RRT). When a new target point emerges mid-mission, the drones reallocate their assigned points. They send images of potential victims, which the human confirms as true victims or misidentified objects due to the drones' limited vision.

The human monitors the drones' navigation, confirms victims, and responds to environmental changes (e.g., wind gusts) via a graphical interface. They also monitor drone positions, altitude, and status and can override the drones to reassign target points or manage harsh wind conditions.

The interface consists of a graphical display, sensors, and an LLM-based feedback module. A Polar H10 heart rate sensor and webcam track the human's physiological data, while a cognitive model estimates their workload in real-time. The LLM-based module provides adaptive mission information by generating natural language feedback through the interface or an auditory text-to-speech (TTS) tool.

Three different fixed function allocation (FA) policies were tested. The functions allocated to the human for each FA are as follows:

- FA 1 (simple): Monitoring the drones' states and confirming identified victims.
- FA 2 (intermediate): All functions in FA 1, plus reallocating drones when a new target point appears.
- FA 3 (complex): All functions in FA 2, plus responding to environmental changes, if any.

In our pilot study, we gathered NASA-TLX results for each FA case, and that reveals that FA 1, FA 2, and FA 3 involve (relatively) low, medium, and high workload.

We have trained a workload model capable of estimating low and high workload levels. The long short-term memory (LSTM) network used for workload estimation includes two LSTM layers with 50 units each, followed by a fully connected layer. The input features include ECG measurements from a heart rate sensor, along with facial landmarks and gaze data from OpenFace [7].

The procedure of the user study is as follows. First, the human is asked to remain in front of the interface to calibrate the sensors. Then, the human performs three sessions of the SAR scenario. In the first session, FA 1 is applied, and the workload is estimated. Once the workload is gathered, the function allocation is adjusted according to the following rule: if the human's workload is high, simplified feedback is provided to reduce the workload, and a less complex function allocation is assigned. Conversely, if the workload is low, the feedback module offers more detailed information to improve transparency, and a more complex function allocation is assigned. Two examples of LLM-generated feedback are shown in Fig. 3.

```
Current workload is low. Overall mission status remains unchanged.

Please perform the following human tasks:
1. Monitor the drone state.
2. Set and update the searching area.
3. Find the victims.

Meanwhile the following drone tasks will be performed by UAV:
1. Monitor the drone faults.
2. Monitor the environmental condition.
```

```
Workload status: High
Warning! Answer will be short and concise to reduce human burden.

1. Monitor the drone state.
2. Set and update the searching area.
3. Find the victims.
```

Fig. 3. The example LLM-generated feedback when the human's workload is low (top) and high (bottom), respectively. A text-to-speech (TTS) tool can generate auditory feedback based on the generated script.

Figure 4 presents the estimated workload of the human. The average workload aligns with the NASA-TLX results for each FA case (FA1, FA2, and FA3). An interesting observation is that high workload was recorded in the latter part of the second session, which was performed with FA2. During this period, target reallocation and victim confirmation occurred simultaneously. This indicates that real-time workload estimation can be used for adaptive function allocation during each session.

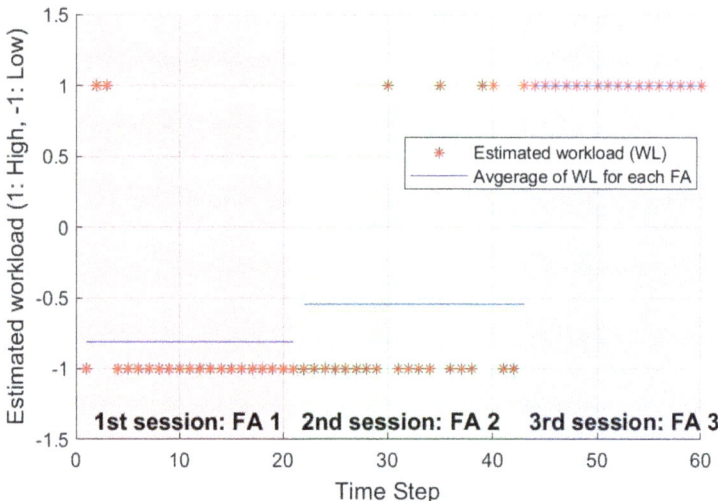

Fig. 4. The estimated workload (WL) for each session and the corresponding allocated function.

8 Conclusions

A computational approach for effective human-swarm collaboration was presented. The key idea of the proposed approach is to explicitly account for the human cognitive state, which can be monitored in real-time using behavioral and physiological data. Our ongoing work can be categorized into three parts. First, we are expanding our cognitive model to incorporate additional cognitive states such as situational awareness and trust. Second, a series of human user studies is being conducted to gather more behavioral and physiological data. Finally, an adaptive function allocation policy (i.e., time-varying policy) and advanced interactive feedback will be developed using the deep Q-learning approach.

Furthermore, our research group has the capability to quickly scale up our initial results using the PURT (Purdue UAS Research and Test facility) with significant expertise in the field of HAC through our work on projects[1] that encompass computational cognitive models [6] and human-autonomy teaming [11]. Based on this, we plan to expand the proposed architecture to more complex and larger scale of platform which several drones and human can interact simultaneously.

Acknowledgments. This work was partially supported by NSF CNS-1836952.

---

[1] Cognitive Autonomy Project (NSF CNS-1836952) and Safe Autonomy Project (NASA 80NSSC20M0160).

References

1. National academies of sciences, engineering, and Medicine: human-AI teaming: state-of-the-art and research needs. The National Academies Press, Washington, DC (2022). https://doi.org/10.17226/26355
2. Tadesse, A.A., Wang, K.J., Lin, C.J.: Task allocation problem between human–robot collaboration team. In: Intelligent and Transformative Production in Pandemic Times, pp. 473–482. Springer (2023)
3. Byeon, S., Choi, J., Zhang, Y., Hwang, I.: Stochastic-skill-level-based shared control for human training in urban air mobility scenario. ACM Trans. Hum.-Robot Interact. **13**(3) (2024). https://doi.org/10.1145/3603194
4. Parasuraman, R., Sheridan, T.B., Wickens, C.D.: Situation awareness, mental workload, and trust in automation: viable, empirically supported cognitive engineering constructs. J. Cogn. Eng. Decis. Making **2**(2), 140–160 (2008). https://doi.org/10.1518/155534308X284417
5. Roth, E.M., Sushereba, C., Militello, L.G., Diiulio, J., Ernst, K.: Function allocation considerations in the era of human autonomy teaming. J. Cogn. Eng. Decis. Making **13**(4), 199–220 (2019). https://doi.org/10.1177/1555343419878038
6. Byeon, S., Choi, J., Hwang, I.: A computational framework for optimal adaptive function allocation in a human-autonomy teaming scenario. IEEE Open J. Control Syst. **3**, 32–44 (2024). https://doi.org/10.1109/OJCSYS.2023.3340034
7. Baltrusaitis, T., Zadeh, A., Lim, Y.C., Morency, L.P.: Openface 2.0: facial behavior analysis toolkit. In: 2018 13th IEEE International Conference on Automatic Face & Gesture Recognition (FG 2018), pp. 59–66 (2018). https://doi.org/10.1109/FG.2018.00019
8. Cardone, D., et al.: Classification of drivers' mental workload levels: comparison of machine learning methods based on ECG and infrared thermal signals. Sensors **22**(19), 7300 (2022)
9. Endsley, M.R.: Toward a theory of situation awareness in dynamic systems. Hum. Factors **37**(1), 32–64 (1995). https://doi.org/10.1518/001872095779049543
10. Chen, J.Y.C., Lakhmani, S.G., Stowers, K., Selkowitz, A.R., Wright, J.L., Barnes, M.: Situation awareness-based agent transparency and human-autonomy teaming effectiveness. Theor. Issues Ergon. Sci. **19**(3), 259–282 (2018). https://doi.org/10.1080/1463922X.2017.1315750
11. Byeon, S., Tian, D., Ayoub, J., Song, M., Moradi Pari, E., Hwang, I.: Optimal function and attention allocation for human-AI collaboration using computational cognition-work model. In: 2024 63rd IEEE Conference on Decision and Control (CDC) (2024). Accepted
12. Cosler, M., Hahn, C., Mendoza, D., Schmitt, F., Trippel, C.: nl2spec: interactively translating unstructured natural language to temporal logics with large language models. In: Computer Aided Verification, pp. 383–396. Springer (2023)
13. Byeon, S., Yuh, M., Choi, J., Jain, N., Hwang, I.: Workload classification for function allocations in human-autonomy teaming using noninvasive measurements. In: AIAA Scitech 2024 Forum, AIAA (2024). Accepted

Open Access This chapter is licensed under the terms of the Creative Commons Attribution 4.0 International License (http://creativecommons.org/licenses/by/4.0/), which permits use, sharing, adaptation, distribution and reproduction in any medium or format, as long as you give appropriate credit to the original author(s) and the source, provide a link to the Creative Commons license and indicate if changes were made.

The images or other third party material in this chapter are included in the chapter's Creative Commons license, unless indicated otherwise in a credit line to the material. If material is not included in the chapter's Creative Commons license and your intended use is not permitted by statutory regulation or exceeds the permitted use, you will need to obtain permission directly from the copyright holder.

Towards GenAI-Empowered Unmanned Traffic Management with Zero Trust

Mohammad Atrouz[✉], Abdulhadi Shoufan, and Fayaz Mohamed Haneefa

Center of Cyber-Physical Systems, Khalifa University, Abu Dhabi, UAE
{mohammad.atrouz,abdulhadi.shoufan,fayaz.haneefa}@ku.ac.ae

Abstract. Unmanned Traffic Management (UTM) systems rely on humans in the loop for reliable and accountable decision-making. Typically, system operators face significant cognitive load due to the high volume of information, the need for multitasking, and real-time problem-solving to ensure safe and efficient airspace management. These demands can lead to cognitive fatigue and an increased risk of errors. Due to their extensive attention abilities, we propose integrating large language models (LLMs) into UTM systems as decision-assistance tools. LLMs can perform real-time data analysis, interpret situations, and suggest optimal solutions, which helps UTM system operators make more reliable, informed, and timely decisions.

This poster shows a use case where we integrate an LLM into an airspace monitoring and surveillance system. This is to verify UAVs' compliance with remote identification regulations and conformance to mission plans to warn the system operator of any violations. We validate the solution by evaluating the accuracy of the LLM responses and the time needed to generate them for different zero-shot prompts in three scenarios. The results show that the LLM can generate helpful messages to the system operator with an accuracy of up to 91.7% within 5.7 s on average. In future work, LLMs will be integrated into other UTM systems. The communication between these LLMs will be enabled and evaluated towards a secured, fully GenAI-empowered UTM with zero trust.

Keywords: UAV · UTM · LLM · Airspace Monitoring and Surveillance · Remote Identification

1 Introduction

Unmanned Traffic Management (UTM) is a system-of-systems that aims to facilitate the safe and efficient integration of Unmanned Aerial Vehicles (UAVs) into the national airspace [1]. It addresses traffic management, real-time communication, and coordination between multiple stakeholders to ensure safe, secure, and private UAV operations. UTM also includes regulations and standards for UAV operations, including UAV registration, flight authorization, and Remote Identification (RID) [2].

© The Author(s) 2026
M. Andreoni and S. Thakkar (Eds.): GENZERO 2024,
Proceedings of 1st GENZERO Workshop, pp. 165–171, 2026.
https://doi.org/10.1007/978-981-95-1050-4_20

UTM promotes data sharing among stakeholders for enhanced situational awareness and decision-making. Due to strict safety and accountability requirements, decision-making in UTM systems is still human-based, which poses many challenges. First, operating a UTM system, such as a surveillance or counter-drone system, requires sustained attention and vigilance since even minor lapses can lead to serious consequences. This constant state of alertness can lead to cognitive fatigue. Moreover, the human operator must handle a large volume of information and manage multiple tasks simultaneously. Additionally, they must communicate with other UTM system operators and address highly complex situations in a dynamic real-time environment. Certainly, these tasks are associated with a high cognitive load that can lead to sub-optimal or wrong decision-making [3].

Generative Artificial Intelligence (GenAI), specifically Large Language Models (LLMs), can improve decision-making processes in UTM systems because they are characterized by extensive attention abilities. LLMs can process large amounts of data and generate human-language messages to assist human operator decision-making.

In this poster, we propose integrating LLMs into UTM systems to assist operators, as illustrated in Fig. 1. We demonstrate this approach for airspace surveillance, as highlighted in the figure.

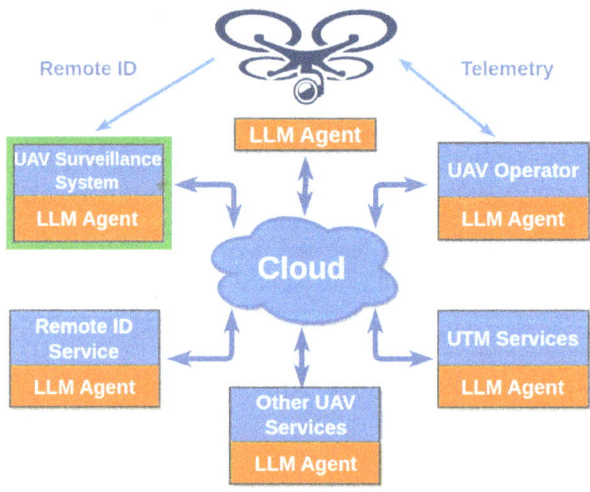

Fig. 1. A hypothetical UTM architecture with possible LLM integrations.

2 Use Case: LLM-Based Airspace Surveillance

Monitoring the airspace against illegal UAV operations is an urgent task to ensure safety and security in national airspace. Detection technologies, including radar, acoustic systems, computer vision, radio-frequency detectors, and remote identification, are essential for airspace monitoring and surveillance. However, these technologies only assist human operators and do not replace them. With the expected increase in UAV operations in the near future, human operators will be overwhelmed by the vast volume of information and data they must process in real-time.

We propose utilizing large language models to enhance airspace monitoring and surveillance systems against malicious drones by supporting human operators in decision-making.

2.1 System Architecture

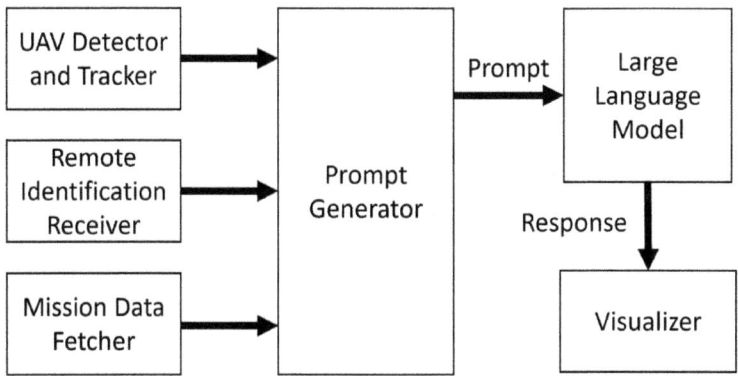

Fig. 2. Architecture of proposed LLM-based airspace surveillance system.

Figure 2 outlines the architecture of the LLM-enhanced surveillance system. The system has a component for drone detection and tracking, e.g., based on computer vision. This component alerts the operator about the existence of a UAV in the detection range. When the drone broadcasts its ID, the Remote Identification Receiver analyzes and authenticates it before forwarding the data to the Prompt Generator. Knowing the drone ID, the system can get information about the mission plan by accessing a dedicated Mission Data Fetcher service.

Based on the data received from the previous modules, the Prompt Generator forms a prompt for the LLM. We define an upper bound for the number of prompts through a parameter we call the *Prompting Interval* that describes the minimum time between two prompts. This parameter aims to control the overhead of the LLM.

2.2 System Operation

Figure 3 illustrates how the LLM-based surveillance system works conceptually. The system only triggers the LLM when a drone is sighted or identified in the detection range. Prompt generation is at the core of this system. The prompt consists of the following sections:

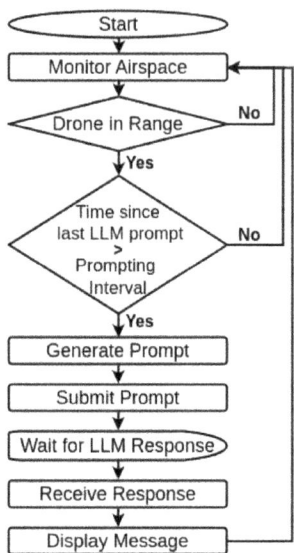

Fig. 3. Operation of the LLM-based Airspace Surveillance System.

1. **Instructions:** In this section, we describe the problem to the LLM, including its role, the geospatial location of the monitoring system, its detection range, the type of data it will receive, and how it should assist. Figure 4 shows a snippet of the prompt.
2. **Dynamic data:** In this section, we provide real-time data such as remote identification messages and mission plans, as shown in the snippet of Fig. 5.
3. **Detailed guidelines (optional):** In this section, we provide more specific information about what needs to be assessed, such as whether the UAV is broadcasting RID, whether it has a published flight plan, and whether the UAV is following this plan within a specified margin of error, among others.

The LLM is prompted to accept and respond in a JSON-structured format; the response is illustrated in Fig. 6.

2.3 Zero Trust

Zero trust is a security framework that operates under the principle of "never trust, always verify," assuming that no user or system-inside or outside the network-should be trusted without continuous verification. This approach requires strict authentication and limits access to only what is necessary for each entity. Modularisation and segmentation help isolate data, nodes, and users within the proposed system. Rigorous authentication and verification steps must precede every interaction to ensure security. The interface to the LLM is especially critical due to the significant volume of data it processes. OpenAI's API, used in this context, implements secure authentication via API keys. Plans

```
LLM_Role = "You are assistant in a UAV surveillance system;
you will be helping the operator to monitor UAV compliance ...
Your location is {..} with a detection range of {..} metres.

You will process the provided data to check UAV compliance,
violations,{..}. Respond with a JSON structure containing
compliance status, risk assessment,{..}.

You must check if the UAV is transmitting Remote ID, risk
assessment of a deviated UAV depends on the context (eg. UAV
is close to no fly zones or not)"
```

Fig. 4. A snippet of the prompt's instruction section.

```
{
    "Recieved_RID_Data": {
        "Flight_id": 23534,
        "Latitude": 24.4549472,
        "Longitude": 54.383438,
        "Altitude_metres": 49.6,
    },
    "Flight_Plan_Server_Data": [ ... ],
    "No_Fly_Zones_Server_Data": [ ... ]
    "Observer_Data": {
      "Visual_Observation": [ ... ],
      "Broadcast_RID_Observation": [..]
    }
}
```

Fig. 5. A snippet of the prompt's data section.

```
{
  "uav_id": 32533,
  "compliance_status_string": "Non-Compliant",
  "violation_type_string": "Outside approved path",
  "risk_level_int": 3,
  "risk_assessment_string": "The UAV is 293 meters away from its
                            approved flight path. It is 1.4 km
                            away from the closest no-fly zone.",

  "suggested_actions_string": "Notify authorities and contact
                              UAV operator"
}
```

Fig. 6. A snippet of the LLM's JSON response.

to test a local LLM will enhance the system's security by keeping the data self-contained and reducing reliance on external services. Furthermore, it is vital that all data, particularly sensitive information such as user or UAV identities, be anonymised before being shared with untrusted entities or nodes that do not require access to this information-such as the LLM model. This includes anonymising and normalising geo-spatial data, which maintains the relationship between data points while concealing sensitive details. These precautions minimise the potential damage from data breaches, ensuring attackers can only access anonymised, non-critical information.

3 Evaluation

We created a software stack to simulate the LLM-based surveillance system.[1] It contains the following components: A *Flight Simulator* for the UAV and the 3D environment; *UTM Server* to simulate backend UTM services needed to support flight plans, Remote ID, and no-fly zones; the *Surveillance System* assuming that an observer can visually detect UAVs and receive Remote ID messages in its range; and an *LLM Instance* that receives prompts from the observer and provides formatted responses. The interface was built with the OpenAI API with our experiments using GPT-4o [4].

Three scenarios were simulated as shown in Fig. 7: A compliant UAV that broadcasts RID and follows the designated flight path, a semi-compliant UAV that broadcasts RID but deviates from its approved flight plan, and a malicious UAV that enters a no-fly zone.

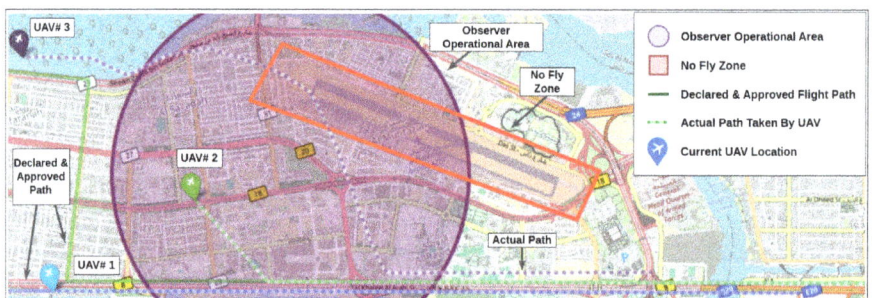

Fig. 7. Simulated scenarios.

Since there are no established systems against which to benchmark, we decided to assess the LLM responses based on their correctness from a human perspective as the baseline. Before the simulation, we defined the expected outputs and the responses from the LLM were compared to evaluate for correctness. Two rounds of testing were performed. In the first round, the LLM received basic

---
[1] Video demo: https://sites.google.com/view/ku-utm/videos/gen-ai-demo.

instructions akin to simple zero-shot prompting. In the second round, we provided detailed guidelines as outlined in Sect. 2.2.

Every scenario was simulated 15 times. With simple prompting, the LLM responses showed an average accuracy of almost 50% across all scenarios. When we added the assessment guidelines, performance significantly improved to 91.7%. Across all of the experiments, the average response time was 5.7 s.

4 Discussion and Conclusion

The preliminary simulation results are promising. However, to better understand the advantages of LLMs, we first need to replicate the scenarios in field experiments with a human observer in the loop and compare the human's performance and response time with those of the LLM.

Future work should also investigate advanced prompting techniques such as chain-of-thought prompting, scenarios involving multiple drones and multiple observers, and increased interactions between observers and other UTM services. Local LLMs should also be considered to improve contextual understanding and reduce response time.

References

1. Hamissi, A., Dhraief, A.: A Survey on the unmanned aircraft system traffic management. ACM Comput. Surv. **56**(3), 1–37 (2023)
2. Belwafi, K., Alkadi, R., Alameri, S. A., Al Hamadi, H., Shoufan, A.: Unmanned aerial vehicles' remote identification: a tutorial and survey. IEEE Access **10**, 87577–87601 (2022)
3. Shoufan, A., Damiani, E.: Contingency clarification protocols for reliable counter-drone operation. IEEE Trans. Aerosp. Electron. Syst. (2023)
4. OpenAI: ChatGPT-4o. https://openai.com/index/hello-gpt-4o/l. Accessed 01 July 2024

Open Access This chapter is licensed under the terms of the Creative Commons Attribution 4.0 International License (http://creativecommons.org/licenses/by/4.0/), which permits use, sharing, adaptation, distribution and reproduction in any medium or format, as long as you give appropriate credit to the original author(s) and the source, provide a link to the Creative Commons license and indicate if changes were made.

The images or other third party material in this chapter are included in the chapter's Creative Commons license, unless indicated otherwise in a credit line to the material. If material is not included in the chapter's Creative Commons license and your intended use is not permitted by statutory regulation or exceeds the permitted use, you will need to obtain permission directly from the copyright holder.

Adaptive Resilient Swarming Using Attention and Reinforcement Learning

Robert Penicka[(✉)] and Martin Saska

Multi-robot Systems Group, Faculty of Electrical Engineering,
Czech Technical University in Prague, Prague, Czech Republic
penicrob@fel.cvut.cz
http://mrs.felk.cvut.cz/

Abstract. In this extended abstract, we introduce a research proposal for a novel learning-based approach to achieve adaptive and resilient swarming of Unmanned Aerial Vehicle (UAV) systems through the integration of attention mechanisms and reinforcement learning. We base it upon our expertise in Reinforcement learning for agile flight in real environments and recent extensions towards learned agile swarming. Our proposed approach will leverage the Transformer network, Variational Autoencoders (VAEs), and Reinforcement Learning (RL) to create an intelligent system capable of real-time adaptation and possible in-flight learning to accommodate reactions to dynamic environments and flight anomalies. The Transformer model processes and analyzes sensor data across the swarm, capturing long-range temporal dependencies to facilitate decision-making and coordination. VAEs provide a compact latent representation of UAV state information, enabling both local anomaly detection and efficient data transmission to edge computing platforms. The RL component learns navigation policies that adapt to the current state of the swarm, guided by insights from both the VAE and Transformer outputs. By combining these technologies, we will achieve a resilient UAV swarm capable of maintaining mission integrity and operational safety even under challenging conditions such as adversarial attacks, communication signal jamming, GPS spoofing, and similar. This research is particularly relevant to participants of the GENZERO workshop, focusing on the intersection of advanced machine learning and autonomous systems.

Keywords: Unmanned Aerial Vehicles · Swarming · Attention · Reinforcement Learning

1 Extended Abstract

Resilience and safety are paramount for swarms of Unmanned Aerial Vehicles (UAVs) as these systems must operate cohesively and reliably in complex and dynamic environments fulfilling various missions. The ability of each drone in the swarm to continuously adapt to changing conditions - such as weather variations, terrain obstacles, evolving mission objectives, or even sensor failure - ensures that the swarm can maintain its operational effectiveness even with encountered anomalies. Fault and threat detection within the swarm is crucial for identifying and isolating compromised or malfunctioning units, preventing cascading failures that could disrupt the entire operation. This continuous monitoring and rapid response capability are essential for maintaining the integrity and safety of the swarm, particularly in high-stakes scenarios such as search and rescue missions, disaster response, or military operations, where the failure of even a single drone could jeopardize the mission's success and safety of all involved. At the same time, the swarm has to adapt to a possibly changing environment while ensuring the safety of its operation concerning the surrounding environment, as well as possible bad actors trying to interfere with the operation, for example through listening to the communications, communication signal jamming, GPS spoofing, and others. Given the recent advancements in machine learning for large sequential data understanding and also learning-based flight, this research proposal plans to employ a learning-based approach for resilient adaptive swarming as illustrated in Fig. 1.

Fig. 1. Our swarm of UAVs controlled with classical methods (top) and agile flight of a single UAV in a cluttered environment (bottom), both illustrating the research proposal idea of learned resilience swarming using a combination of attention mechanism and reinforcement learning.

In particular, machine learning models like Transformers [1], Variational Autoencoders (VAEs) [2], and Reinforcement Learning (RL) [3] each play a pivotal role in advancing the capabilities in various fields ranging from natural language processing and dimensionality reduction to autonomous systems' advanced autonomy. Transformers, with their ability to capture long-range temporal dependencies and relationships within data, can enable more effective processing and analysis of the vast amounts of information generated by drones within a swarm. This capability allows for real-time interpretation of sensory data, facilitating smarter and faster decision-making for adaptation of the swarm in dynamic situations or during anomaly situations. VAEs can contribute by efficiently encoding high-dimensional UAV sensory data captured onboard and from external localization sources into lower-dimensional latent space while making it possible to detect anomalies within the data. This is crucial for maintaining the operational integrity of the drones, as VAEs can identify potential issues before they escalate into critical failures. Reinforcement Learning (RL) complements these technologies by enabling drones to learn optimal behaviors through trial and error in simulated environments, allowing them to adapt to new challenges and refine their strategies over time. At the same time, the RL swarm navigation policy can leverage even the embedded VAE-encoded data and the outcomes of the transformer-based analysis of the swarm, and use both to adapt the swarm navigation and thus react to changing environment and state of the swarm. Together, Transformers, VAEs, and RL form a powerful combination that can make the swarm more intelligent, adaptive, and secure.

We overview the proposed system in the following subsections.

Variational Autoencoder for State Representation and Local Anomaly Detection. The Variational Autoencoder (VAE) will be designed to create a compact latent representation of the drone's state, incorporating all sensor data, including the relative localization of other drones in the swarm, incoming communication data, control deviations, and control inputs. The autoencoder reduces the dimensionality of this information while preserving the essential details about the UAV in a much smaller latent space. The VAE serves a dual purpose in the proposed learning-based adaptive resilient swarming approach. First, its latent space representation can be used for local onboard safety by classifying the embeddings into categories of possible anomalies, such as critical, major, moderate, and minor. The encoder of the VAE can act as a feature extraction tool for traditional classification methods like k-Nearest Neighbors, Logistic Regression, or Decision Trees. This classification system can be used not only to identify the severity of anomalies but also to detect specific cases, such as sensor failures or fraudulent communication data. Because the VAE clusters similar input data closely in the latent space, it enables the classification system to detect anomalies that were not specifically accounted for during training but exhibit similar symptoms, allowing for effective resolution. Additionally, the mere existence of an anomaly can be detected by identifying outliers within the normal distributions represented in the VAE's latent space, thereby enhancing local, single-UAV onboard resilience. The second purpose of the VAE is to create a compact representation of the UAV's state and sensor data that can be transmitted to an edge base station computer. The latent data generated by the VAE's encoder are significantly smaller, retaining the essential information while being suitable for transmission over low-bandwidth communication channels. This compressed representation is also secure, as the data cannot be interpreted by adversaries without access to the VAE's learned decoder. Finally, the embedded latent representation can be utilized by a learned navigation policy within the swarm to adapt its behavior dynamically.

Transformer Model for Swarm Adaptation and Status Monitoring. Utilizing Transformer models in drone swarms can significantly enhance the processing and analysis of data collected from individual drones regarding their operational states. Transformers, known for their ability to capture long-range dependencies and relationships within data, can effectively receive embeddings from Variational Autoencoders (VAEs), which compress complex sensor data into concise, informative representations. By analyzing this temporal data, Transformers can identify patterns, detect anomalies, understand environmental changes, and predict potential issues within the swarm, thereby serving as a crucial component for global decision-making and collective safety mechanisms. The powerful attention mechanisms within Transformers allow them to focus on the most critical aspects of the data, facilitating real-time decision-making and coordination across the swarm. The Transformer model can either run on the UAV or on a base station, receiving low-bandwidth data from the VAE from all UAVs through various communication channels. As the Transformer network processes the entire history of encoded data from the VAEs about the sensors and states of all UAVs in the swarm, it can identify anomalies related to mission objectives, the overall state of the swarm, individual UAVs, and specific sensors. In addition to anomaly detection, Transformers play a vital role in influencing the behavior of the swarm. The output of the Transformer is not only capable of generating textual descriptions of the swarm's state but more effectively produces an embedded information vector that can be integrated into UAV navigation policies. This vector can encapsulate an analyzed situation regarding the entire swarm, the mission, and the individual UAVs, and can thus trigger behavioral adjustments such as landing a drone with a failed sensor or altering the swarm's movement in response to environmental changes. By doing so, the Transformer optimizes swarm behavior, enhances resilience, and improves performance in complex and dynamic environments, all while reducing the need for extensive manual oversight.

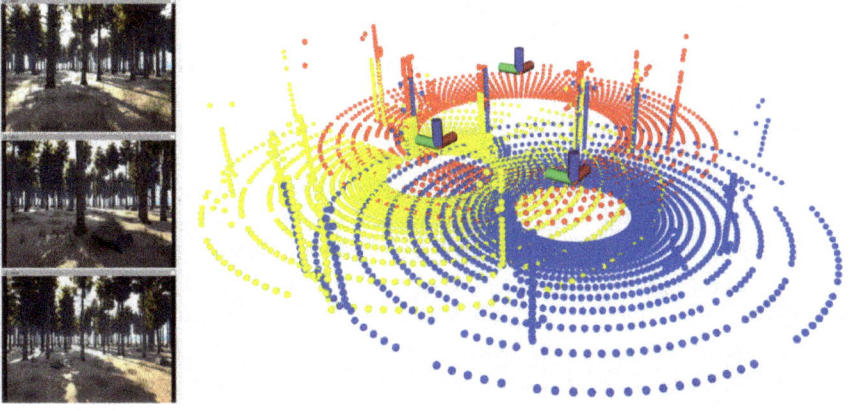

Fig. 2. Showcase of a swarm simulation with 3 UAVs in Unreal Engine 5.

Reinforcement Learning for Adaptive Swarm Navigation (RL). We plan to use Reinforcement Learning (RL) [4] for the navigation of UAV swarms, building on our previous work in [5]. However, unlike in [5], where RL was used as a control policy, we will now employ RL as a navigation policy. This shift simplifies the generated actions, allowing the policy to command existing UAV controllers and thus make the learning of generalizing policy easier. Additionally, we intend to deploy this navigation policy on real UAVs, utilizing our UVDAR system [6] for the relative localization of the UAVs within the swarm. The observation space of the swarming flight policy will include not only the UAV's state and the relative positions of other UAVs in the swarm but also the latent representations from the VAE and the output from the Transformer. These embedded data will provide comprehensive information about the entire swarm. Incorporating the VAE latent data allows the navigation policy to respond effectively to local anomalies and threats, such as guiding a UAV with a failed sensor to land safely or relying more on IMU and relative localization data when GPS data is unreliable. Furthermore, the Transformer's data will enable the UAV navigation policy to adapt based on the state of the entire swarm as analyzed by the Transformer. This capability allows the swarm to change its behavior in response to alterations in the environment or mission objectives. We chose Reinforcement Learning for navigation over more traditional planning pipelines because RL can directly incorporate data from both the VAE and the Transformer, enabling the navigation policy to learn how to adapt to the environment and complete missions with resilience to potential anomalies.

Training. We plan to conduct the training of our UAV swarm navigation system using our custom Unreal Engine 5 (UE5) simulator (shown in Fig. 2), which offers a highly dynamic and adaptive environment. This simulator is capable of simulating a variety of conditions, including changing weather, system failures, and diverse environmental challenges, allowing us to thoroughly train and test the swarm's resilience in a wide range of scenarios. The Variational Autoencoder (VAE) will be trained independently to process all data from the individual UAVs, enabling it to effectively compress and analyze sensor information. We will also train the VAE to classify a specified set of anomalies, ensuring that the system can detect and respond to common issues effectively. In parallel, the Reinforcement Learning (RL) navigation policy will be trained together with the Transformer model. This joint training will enable the RL policy to learn how to interpret the embedded information about the swarm produced by the Transformer.

Additionally, our approach can include the capability for continual learning during deployment. The VAE can detect anomalies that may indicate the UAV is operating in conditions outside the range of the data it was originally trained on, such as in a different environment or a new security thread. When such anomalies are detected, the RL policy can continue to learn during flight and include the new conditions, adapting to new data and further enhancing the UAVs' ability to navigate and respond to unexpected challenges. This ensures that the UAVs learn to navigate and adapt to real-world conditions, building the resilience to handle unexpected challenges both during training and in live operations.

2 Conclusion

In this research proposal extended abstract, we presented a framework for adaptive and resilient UAV swarming using a combination of Variational Autoencoders (VAEs), Transformer models, and Reinforcement Learning (RL). Building on our expertise in RL for agile flight, we plan to extend our proven single-drone navigation techniques to multi-drone swarms, focusing on real-world applications where UAVs must adapt to dynamic environments, possible security threads, and operational safety anomalies. Our existing work on RL navigation serves as a foundation as we plan to integrate VAE for anomaly detection and Transformers for adaptive swarm coordination. This positions our research proposal to advance the state-of-the-art in UAV swarm autonomy.

References

1. Vaswani, A.: Attention is all you need. arXiv preprint arXiv:1706.03762 (2017)
2. Kingma, D.P., Welling, M.: Auto-encoding variational bayes. arXiv preprint arXiv:1312.6114 (2014)
3. Sutton, R.S., Barto, A.G.: Reinforcement Learning: An Introduction, 2nd. MIT Press (2018). http://incompleteideas.net/book/the-book-2nd.html
4. Penicka, R., Song, Y., Kaufmann, E., Scaramuzza, D.: Learning minimum- time ight in cluttered environments. IEEE Robot. Autom. Lett. **7**(3), 7209–7216 (2022)
5. Poncar, K.: Reinforcement learning for swarm control of unmanned aerial vehicles. BA thesis, Czech Technical University in Prague (2023)
6. Walter, V., Staub, N., Franchi, A., Saska, M.: Uvdar system for visual relative localization with application to leaderfollower formations of multi- rotor UAVs. IEEE Robot. Autom. Lett. **4**(3), 2637–2644 (2019)

Open Access This chapter is licensed under the terms of the Creative Commons Attribution 4.0 International License (http://creativecommons.org/licenses/by/4.0/), which permits use, sharing, adaptation, distribution and reproduction in any medium or format, as long as you give appropriate credit to the original author(s) and the source, provide a link to the Creative Commons license and indicate if changes were made.

The images or other third party material in this chapter are included in the chapter's Creative Commons license, unless indicated otherwise in a credit line to the material. If material is not included in the chapter's Creative Commons license and your intended use is not permitted by statutory regulation or exceeds the permitted use, you will need to obtain permission directly from the copyright holder.

Challenge 6: – Cloud Security and Privacy

Protecting the Intellectual Property of QNNs at the Deep Edge

Miguel Costa[✉], Tiago Gomes, and Sandro Pinto

Zero-Day Labs, Universidade do Minho, Braga, Portugal
miguel.costa@dei.uminho.pt

Abstract. This study presents the first framework designed to protect the intellectual property of Quantized Neural Networks (QNNs) on Arm Cortex-M MCUs equipped with TrustZone-M. Our framework - SecureQNN - employs an iterative simulation to identify the most privacy-critical layers of a QNN, strategically offloading their execution to the secure-world of TrustZone-M. For each set of private layers, SecureQNN computes the epochs an adversary should employ to reconstruct a substitute QNN that is at least as accurate as the target QNN. The set of layers allowing the adversary to train with less effort than the original training process is delegated to the secure-world of TrustZone-M. Results for QNNs trained on CIFAR-10 and Visual Wake Words (VWW) datasets suggest that it is possible to increase the privacy of a QNN by delegating only 51% to 65% of the total model size to the secure-world.

Keywords: QNNs · Arm Cortex-M · TrustZone-M · Privacy

1 Introduction

Developing efficient Machine Learning (ML) models usually requires years of expertise. Therefore, this task has been mostly performed by big data companies that sell their models in the form of Machine Learning as a Service (MLaaS). However, traditional MLaaS follows a centralized computing paradigm, requiring transmitting private client data to cloud servers for inference. The rising concerns about data privacy aligned with the lack of adaptability of MLaaS for real-time or mission-critical scenarios, due to the unpredictable network and cloud latency, are driving a paradigm shift toward ML at the deep edge of the network [1].

However, deploying proprietary ML models in untrusted devices is not often in the interest of the Service Provider (SP) as it raises concerns about unauthorized access. Such exposure can result in the reverse engineering of proprietary algorithms, leakage of training data, or contribute to the use of white-box adversarial attacks to compromise decision integrity [1]. A promising approach to mitigating these risks involves using Trusted Execution Environments (TEEs). As a result, there has been a growing interest among researchers in modifying commercially off-the-shelf TEEs - ARM TrustZone and Intel-SGX - to protect the intellectual property of ML models. However, TEEs are optimized for executing small-scale security-critical tasks, challenging their suitability for resource-intensive ML workloads.

Although significant progress has been made in the domain of ML model privacy in Arm Cortex-A and x86 application processors (APUs), previous research neglects Arm Cortex-M MCUs, which power most IoT applications. This paper introduces the first framework to protect the intellectual property of Quantized Neural Networks (QNNs) on Arm Cortex-M MCUs equipped with TrustZone-M. The proposed framework simulates a powerful and iterative attack to identify the most privacy-critical layers of a given QNN, delegating their execution to the secure-world of TrustZone-M. With effect, the framework assumes a worst-case threat-model: the adversary can access and manipulate the full software stack of the normal-world and leverage all available information to reconstruct the model. Our framework attests to the intellectual property of a QNN if the minimum effort/computational cost required to reconstruct the QNN is at least equivalent to the computational cost incurred by the SP in training the original QNN from scratch, measured in training epochs. Experimental results on CIFAR-10 and Visual Wake Words (VWW) datasets indicate that privacy can be preserved by offloading only 51% to 65% of the QNN size to the secure-world.

2 Model Partitioning

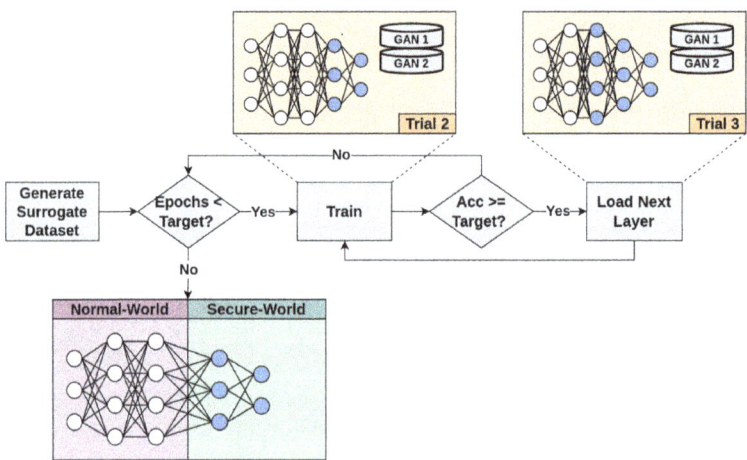

Fig. 1. Model partitioning workflow: evaluation of the criticality of the last two layers.

Yosinski et al. [2] proved that early layers of Artificial Neural Networks (ANNs) play a fundamental role in feature extraction, whereas the deeper layers are more crucial for classification. Our framework leverages this insight to partition a given QNN into two contiguous blocks: one set of public layers assigned to the normal-world, and one set of private layers assigned to the secure-world. We believe this is a promising approach to minimize the TEE memory footprint, as later layers usually feature smaller feature maps. The partitioning process is illustrated in Fig. 1. The framework is based on TensorFlow v2 API, which enables seamless quantization of floating-point ANNs to integer-precision QNNs, following a quantization policy compatible with TensorFlow Lite Micro and CMSIS-NN - the state-of-the-art ML libraries for Arm Cortex-M MCUs [1].

SecureQNN analyzes the QNN progressively, starting at the last layer and moving backward. At each stage, the weights and bias (parameters) of current and following layers are considered unknown to the attacker and set as trainable. The others are considered known to the attacker and, therefore, set as non-trainable. For each set of unknown layers, the framework simulates a powerful training session to determine the minimum epochs necessary for reconstructing a QNN with an accuracy that matches or exceeds the original QNN. The smallest set of unknown layers that requires the attacker paying, at least, the same number of training epochs required to train the source/target model from scratch is assigned to the secure-world. SecureQNN operates under a powerful threat model, assuming the adversary knows the full QNN architecture and hyper-parameters of the original training procedure.

Since attackers do not have access to the original training data, they rely on alternative data sources to approximate the original training conditions. Our framework simulates this scenario by generating a surrogate dataset through advanced data augmentation techniques, based on Generative Adversarial Networks (GANs), applied to the test subset of the original dataset.

3 Generation of a Surrogate Training Dataset

Deep Generative Model Architecture. To identify the most suitable generative algorithm for our use case, we referred to a study conducted in 2022 [3], which benchmarked: (i) Generative Adversarial Networks (GANs), (ii) Energy-Based Models (EBMs), (iii) Variational Autoencoders, (iv) Autoregressive Models, and (v) Normalizing Flows. Results indicate that GANs and EBMs significantly outperform the other architectures. Ultimately, we selected GANs due to the availability of a Python library, StudioGAN [4], which facilitates the experimentation with state-of-the-art GANs.

GAN Evaluation Backbone. The starting point for the GAN design was the evaluation backbone. The evaluation backbone is a pre-trained neural network used to extract a feature set from synthetic and real images to compute the metrics that quantify the quality of the images generated by the GAN. StudioGAN benchmarks three different evaluation backbones: (i) TensorFlow-InceptionV3, (ii) PyTorch-SwAV, and (iii) PyTorch-Swin-T. We observed that the PyTorch-SwAV held better results - Fréchet Inception Distance (FID) - for the CIFAR-10 and ImageNet datasets.

CIFAR-10. As the attacker only has access to a small dataset (the original test dataset only contains 10k samples), we need to apply techniques for data-efficient training, otherwise, the discriminator memorizes the training dataset and makes the training collapse. From the set of GANs implemented in StudioGAN, we only considered options with data augmentation techniques or LeCam regularization. The last design parameter to consider is the training backbone, i.e. the architecture of the generator and discriminator (GAN architecture). Whenever possible, for the same dataset we select two distinct GAN architectures to increase the diversity of the generated dataset. For CIFAR-10, StudioGAN benchmarks two GAN architectures coupled with efficient training and PyTorch-SwAV as the evaluation backbone: (i) StyleGAN and (ii) BigGAN. Regarding the former, StyleGAN2 with Differential Augmentation holds the best FID score in the benchmark and we got good quality images from it (FID = 0.91). Regarding the latter, BigGAN with LeCam regularization holds the best FID score. Unfortunately, LeCam regularization was not sufficient to avoid overfiting. By replacing LeCam regularization with Differential Augmentation we could generate good quality images (FID = 1.59).

Visual Wake Words (VWW). As StudioGAN does not benchmark GANs for the VWW dataset or its resolution (96 x 96), we adapted configurations from other datasets and applied a bicubic filter for image resizing. Among the image resolutions supported by StudioGAN, 128 x 128 is the closest, corresponding to the resolution of the ImageNet dataset. Within ImageNet, StudioGAN benchmarks the BigGAN and StyleGAN architectures. Regarding the former, ReACGAN and BigGAN-256 configurations benchmark the best FID scores. However, both configurations make the training collapse on the VWW dataset. Regarding the latter, StyleGAN2 and StyleGAN3 produce higher-quality images, achieving a FID of 4.17 and 7.05, respectively. We included these two GANs in our framework to simulate a surrogate VWW dataset crafted by a possible attacker.

4 Trusted Execution

The architectural layout of an edge device executing a QNN deployed via SecureQNN is depicted in Fig. 2. To establish two isolated execution environments, we employ TrustZone-M, which partitions the system into (i) normal and (ii) secure worlds. These environments support privileged and non-privileged execution levels. As shown in Fig. 2, privileged components are represented in gray, whereas non-privileged elements are distinguished by color - green for the secure-world and red for the normal-world. This architecture is designed to protect against software side-channel attacks. For now, SecureQNN does not provide guarantees against physical side-channel attacks.

In the privileged level of the secure-world, Secure Boot ensures the integrity of the firmware image, including the private and public layers of the QNN. During the boot process, the system verifies the firmware's digital signature to detect unauthorized modifications. If tampering is detected, the boot sequence halts. SecureQNN uses Trusted Firmware-M (TF-M) as a reference implementation for secure firmware. While the Firmware Update is optional, it can be beneficial if

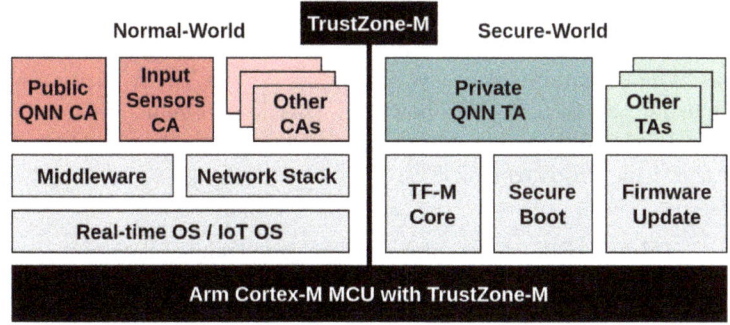

Fig. 2. High-level architecture of an edge device using SecureQNN.

the SP needs to update the QNN remotely. This functionality, also managed by TF-M, ensures that only firmware with a valid digital signature is installed.

In the non-privileged level of the secure-world, the only mandatory Trusted Application (TA) is the Private QNN, which performs the computation of the private QNN layers. Anytime this TA is called, it loads the private layer parameters from a secure memory region and uses CMSIS-NN to calculate the output of the set of private layers. The predicted class, computed in the last QNN layer, is then transmitted to the Public QNN Client Application (CA).

In the normal-world, the only mandatory CAs are the Input Sensors and the Public QNN. The Input Sensors CA manages the sensor or network system to gather the QNN input data. The Public QNN CA starts and manages the inference till the first private layer appears on the pipeline.

5 Results

We evaluate SecureQNN using QNNs from the MLPerf Tiny benchmark [5], as these QNNs reflect the use cases described by more than 30 entities from industry and academia for TinyML systems. We focus on image applications, as images offer a practical means to assess the quality of the surrogate training dataset generated by our framework. Regarding image processing applications, MLPerf Tiny comprises two QNN architectures: (i) RestNet trained for CIFAR-10 and (ii) MobileNetV1 trained for VWW. CIFAR-10 is composed of 32 x 32 images distributed over 10 classes representing various objects. VWW is a binary dataset composed of 96 x 96 images that represent the use case of detecting the presence of a person in an image. Larger ML models, such as GANs or Large Language Models (LLMs), are excluded from the scope of SecureQNN due to the memory and computational limitations of Arm Cortex-M MCUs.

5.1 Surrogate Training Dataset

Table 1 reports the FID of the GANs included in our framework. Additionally, it reports the accuracy of the QNNs when evaluated on the images generated by these GANs and compares these results to the baseline test accuracy. We observe that the selected GANs produce images that closely resemble the original dataset. The maximum accuracy loss occurs for the VWW surrogate dataset generated using the StyleGAN2; however, this loss remains relatively low at 4.11%. For CIFAR-10, the maximum accuracy loss is also in this GAN architecture, but the value drops to 1.53%.

Table 1. FID and accuracy of surrogate training datasets

Dataset	GAN	Input Dim	FID	Accuracy	
				Surrogate	*Original*
CIFAR-10	StyleGAN2+DiffAug	32x32	0.91	85.00%	86.53%
	BigGAN+DiffAug	32x32	1.59	88.38%	
VWW	StyleGAN2	128x128	4.17	81.77%	85.88%
	StyleGAN3	128x128	7.05	83.21%	

5.2 Memory Footprint

Results suggest that it is possible to preserve the privacy of the CIFAR-10 QNN by only delegating the last 3 out of 10 layers to the secure-world. In practice, this corresponds to port 50.93% of the total QNN size to the secure-world, imposing a minimum TEE memory footprint of 39.16 KBytes. To protect the privacy of the VWW QNN, we need to delegate the last 9 out of 28 layers to the secure-world. This corresponds to 65.27% of the total QNN size, which demands a minimum TEE memory footprint of 139.63 KBytes.

5.3 Latency

Our framework implies two context switches per sample between the normal and secure worlds. However, performance measurements on the STM Nucleo-L552ZE-Q board, which features a single Arm Cortex-M33 core at 110 MHz, indicate that the efficiency gained by dividing the execution of QNNs into two parts - thereby reducing intermediate buffer sizes, among other benefits - compensates for this overhead. When the execution of the QNNs is fully delegated to the normal-world, we measure a decision latency per sample of 111.45M and 88.16M clock cycles (1.01 s and 0.8 s @ 110 MHz) for CIFAR-10 and VWW, respectively. After porting the critical layers to the secure-world, these values reduce to 111.44M and 88.04M (1.01 s and 0.8 s @ 110 MHz).

6 Conclusion and Roadmap

This work introduces the first framework designed to protect the intellectual property of QNNs on Arm Cortex-M MCUs supporting TrustZone-M technology.

Despite being out of scope, the model partitioning strategy is valid to protect an ANN/QNN in more advanced SoCs, such as those featuring ML accelerators, as long as the SoC supports TrustZone or similar technology. In such a scenario, the critical layers would be executed within the TEE, while the others could be delegated to the ML accelerator. As a long-term goal, we plan to upgrade our framework to evaluate the privacy of non-consecutive sets of layers as a mechanism for reducing the TEE memory footprint even more.

References

1. Costa, M., et al.: David and Goliath: an empirical evaluation of attacks and defenses for QNNs at the deep edge. In: IEEE EuroS&P (2024)
2. Yosinski, J., et al.: How transferable are features in deep neural networks?. In: International Conference on Neural Information Processing Systems - Volume 2 (2014)
3. Bond-Taylor, S., et al.: Deep generative modelling: a comparative review of VAEs, GANs, normalizing flows, energy-based and autoregressive models. IEEE Trans. Pattern Anal. Mach. Intell. **44**, 7327–7347 (2022)
4. Kang, M., et al.: StudioGAN: a taxonomy and benchmark of GANs for image synthesis. In: IEEE Trans. Pattern Anal. Mach. Intell. **45**, 15725–15742 (2023)
5. Banbury, C., et al.: MLPerf tiny benchmark. In: Proceedings of the Neural Information Processing Systems Track on Datasets and Benchmarks (2021)

Open Access This chapter is licensed under the terms of the Creative Commons Attribution 4.0 International License (http://creativecommons.org/licenses/by/4.0/), which permits use, sharing, adaptation, distribution and reproduction in any medium or format, as long as you give appropriate credit to the original author(s) and the source, provide a link to the Creative Commons license and indicate if changes were made.

The images or other third party material in this chapter are included in the chapter's Creative Commons license, unless indicated otherwise in a credit line to the material. If material is not included in the chapter's Creative Commons license and your intended use is not permitted by statutory regulation or exceeds the permitted use, you will need to obtain permission directly from the copyright holder.

A Zero-Trust Hardware Platform for LLMs and Generative AI Edge Applications

Francesco Restuccia[1]($\boxtimes$), Davide Rossi[2], and Ryan Kastner[1]

[1] University of California San Diego, San Diego, CA, USA
{frestuccia,kastner}@ucsd.edu
[2] University of Bologna, Bologna, Italy
davide.rossi@unibo.it

Abstract. Generative AI and LLMs require significant computing performance. When used in critical applications like UAVs, UGVs, and other edge devices, they additionally need strong safety and security assurances. A zero-trust SoC hardware platform is crucial for delivering the necessary performance to execute advanced ML models while ensuring safe operation and securing data communication and computation. We propose a zero-trust SoC platform to support the deployment and operation of generative AI and LLMs based on (i) a zero-trust system interconnect, (ii) hardware-level fine-grained real-time analysis, (iii) fault-tolerance in on-chip communications and computation; and (iv) advanced mitigations for side-channel attacks. Our proposal is based on several research projects published in top-notch conferences and journals, a collaboration in a relevant SoC project, and collaborations with relevant industrial partners. Our proposal is based on research projects published in top-notch conferences and journals, a collaboration in a relevant SoC project, and partnerships with relevant industrial partners.

Keywords: System-on-Chip · Zero-trust services · Real-time analysis · Fault tolerance · Side-channel mitigations

1 Research Problem

Efficiently computing modern Generative AI and LLM applications on the edge mandates the development of complex system-on-chip (SoC) platforms with strict performance, security, and safety requirements. Given their complexity and to enable fast development, modern SoCs are typically composed of third-party hardware IPs. For example, Al Saqr [1–3] uses CVA6 RISC-V microprocessors, a PULP parallel compute cluster, an OpenTitan hardware root of trust, and various open-source on-chip interconnect and I/O engines that must coordinate to execute safely, securely, and efficiently. Specialized IPs come from different companies and institutions with diverse security and safety approaches.

Hardware threats are extremely dangerous in critical systems. For instance, a single misbehaving IP easily disrupts the communication of the entire SoC, even though the IP adheres correctly to the industry standard on-chip communication protocol [4–6]. The disruption is due to a lack of protocol specification and provides a means to launch hardware and software security attacks capable of disrupting the whole system execution. Disruptions are unacceptable in SoCs designed for critical applications. *A zero-trust paradigm that grants no implicit trust to compute resources and data is crucial for safe and secure operation.* If the SoC is not safe and secure, the operations that run upon it cannot be trusted.

We propose a zero-trust hardware SoC platform targeting the requirements of modern LLMs and generative AI applications. The key aspects are: (i) a *zero-trust system interconnect* (Sect. 2.1) providing advanced services for enforcing safe and secure on-chip communications and system-level monitoring and control features. These services have been demonstrated in several works to be foundational for the secure and safe execution of the system [7,8]; (ii) a *hardware-level real-time analysis and predictability enhancements* (Sect. 2.2), to provide strong timing predictability guarantees and a method for the proper configuration of the platform in applications with hard real-time constraints such as autonomous vehicles and robotics [9]; (iii) *hardware fault tolerance* to promptly detect and correct errors in on-chip communications and investigate the sensitivity neural network models to faults. These are especially fundamental in memory-intense tasks, such as modern AI applications (Sect. 2.3); (iv) *side-channel attacks* mitigations [10] (Sect. 2.4), to secure the system against unintended data leakage. We will implement these methodologies into the Al Saqr SoC [1–3] or for the development of other next-generation SoCs of interest targeting security- and safety-critical applications.

Our team has complementary expertise in hardware security, RTL design, SoC integration, on-chip communications, hardware-enabled safety, and real-time analysis. We have recent successful collaborations presented in this proposal, including a published joint paper in a top-notch conference [8], a joint journal paper in a flagship journal [9], and an ongoing collaboration on the Carfield heterogeneous SoC research project for automotive applications [7]. Carfield has recently been taped out in a joint academic and industrial collaboration led by ETH Zurich and the University of Bologna. The methodologies presented in this proposal enforce security and safety in the Carfield SoC.

2 Methodology

We will develop zero-trust SoC mechanisms and services focusing on safe and secure interconnect, real-time and predictable system analysis, fault tolerance, and side-channel mitigations. We implement these mechanisms and analysis on open-source SoC designs, including Carfield and the Al Saqr drone SoCs. Our mechanisms have a clear TRL4 Proof of Concept and are silicon-proven.

2.1 Zero-Trust System Interconnect

Secure and reliable on-chip communication is a vital part of any zero-trust SoC. It ensures that data firmware is loaded correctly, critical computing assets and data are accessed appropriately, and data transfers are safely and securely executed.

The system interconnect uses on-chip communication standards, such as the industry-standard AMBA AXI [11]. The AXI protocol is not fully specified to facilitate designer flexibility and increase performance. But this creates severe safety and security threats – a single IP can influence the execution of other critical SoC resources by stealing bandwidth [4] or performing a Denial-of-service (DoS) for shared resources [5,6]. These hardware threats are particularly dangerous when using shared resources, which is common in modern SoCs. A malicious/misbehaving module can delay the execution of a critical computation, disrupting the execution of critical tasks, which can have potentially catastrophic consequences. Modern AI and LLM applications have huge memory requirements and are often memory-bound. When used in safety-critical situations, missing a communication deadline can be catastrophic, causing a drone or a car to crash. We developed many zero-trust SoC mechanisms to analyze and address these threats [4–6,8,12]. AXI-Realm [8] deploys a set of mechanisms for predictably managing the access to shared resources in an optimized design for high performance, adding only one cycle of latency and extremely low area overhead. The Carfield SoC [7] uses AXI-Realm.

We propose a *zero-trust system interconnect* as a foundational component for SoC security and safety and removing the implicit trust currently required by the AXI standard. We build upon AXI-Realm [8] to create a zero-trust on-chip interconnect providing (i) high-performance data transfers, (ii) low area impact, and (iii) full compliance with the industrial on-chip communication standard for the full compatibility with existing IPs, such as RISC-V and ARM processors, and ML accelerators. Our zero-trust interconnect enhances the AXI-Realm services [8] is used in Carfield [7] to enforce predictable and controlled on-chip resource access using safety monitors. It provides services like (i) proactive detection and mitigation of attacks/misbehavior generated at software or hardware levels and aimed at bandwidth-stealing [4], unpredictable delays, and DoS of shared resources [5,6]; (ii) predictable fine-grained runtime bandwidth redistribution [12]; (iii) provisioning of statistics including transactions' response times and number of transactions issued to each peripheral; (iv) notification to the HWRoT when security attacks or unsafe conditions are detected.

Our zero-trust system interconnect is integrated with the OpenTitan HWRoT [13] for its runtime configuration and management. Communications between the HWRoT and the interconnect happen via a dedicated secure bus or using authentication. Our zero-trust interconnect is an active extension of the HWRoT, providing features like runtime bandwidth redistribution among the controllers, temporary isolation of one or more controllers from accessing the system to adapt to the criticality of the executed tasks or the platform's current needs, and detection of unsafe conditions and security attacks to spark a prompt defensive response of the system. Our fine-grained real-time analysis provides a methodology to support the configuration to satisfy a specification of a target critical application (see Sect. 2.2).

2.2 Real-Time Analysis and Predictability Enhancements

We developed many zero-trust analysis and modeling techniques for AXI-based network infrastructures [14,15]. Real-time analyses are generally based on the documentation provided by the vendors. However, these analyses intrinsically depend on the quality of the documentation provided and thus are generally limited in precision. This leads to over-provisioning and pessimism on one end and unsafe analysis on the other, possibly leading to deadline misses. We developed an innovative approach to real-time analysis based on the direct modeling and analysis of the RTL describing the hardware architecture to provide a fine-grained holistic analysis building directly from the hardware description language. We applied this methodology to model, analyze, and upper bound the execution of critical tasks on a Cheshire-based open-source SoC in a recent collaboration involving the UC San Diego, the University of Bologna, and ETH Zurich. Our methodology shows significantly improved precision, leading to considerably reduced pessimism w.r.t. previous works [9]

A fundamental requirement in critical SoCs is ensuring timing predictability, especially in applications with hard real-time constraints. We propose to develop and extend our analysis methodologies to SoC platforms of interest, such as Al Saqr, and provide strong timing guarantees for the critical tasks executing on the platform. We extend our analysis to integrate the credit-based bandwidth managers and other zero-trust services introduced in Sect. 2.1. In addition to providing strong real-time upper bounds, the analysis gives a methodology to configure the zero-trust services safely. The zero-trust system interconnect provides a critical workload specification to execute on the platform. Our real-time analysis approach reduces pessimism by safely obtaining higher platform performance, i.e., ensuring that the critical tasks executed on top of the platforms can be completed within their deadline.

We propose integrating a shared scratchpad memory peripheral (DCSPM) into the SoC platform as a key enabler for predictable temporary data access. DCSPM can be dynamically configured between contiguous and interleaved addressing modes at runtime. Contiguous address mapping enables bank privatization within the DCSPM, isolating tasks that should not interfere and simplifying the computation of worst-case delay bounds for real-time analysis (see Sect. 2.2). Interleaved address mapping is beneficial when two tasks with different criticalities share data, as it distributes access over all banks, mitigating conflicts.

2.3 Hardware Fault Tolerance

Ensuring continuous and reliable operation is paramount in critical edge applications, even in the presence of transient faults. Robust fault tolerance mechanisms enhance the system's resilience and align with the principle of minimizing trust and safeguarding critical task execution. Providing prompt error detection and correction features in communications is a foundational feature for zero-trust SoCs, especially considering the enormous memory footprint of modern generative AI and LLM applications requiring continuous data movement – a fault not promptly detected can generate extremely dangerous conditions capable of compromising the system integrity, potentially leading to catastrophic consequences. Most commercial and open-source system interconnects are not conceived with fault tolerance features. We investigate the tradeoff in performance and area and develop and integrate high-performance mechanisms for fault tolerance in on-chip communications for modern SoC platforms of interest. We will consider the most popular approaches: (i) detect and resubmit; and (ii) ECC correction, and investigate their trade-offs [16–18]: in (i), extra area is mandated for data buffering for possible re-submissions while latency is impacted only when a re-submission is requested; in (ii), extra combinatorial area is required for deploying the ECC algorithm while latency is always impacted by computing the ECC algorithm. This enhances the safety and security of the system in the presence of faults in communications.

Besides on-chip communications, we investigated fault tolerance in neural network (NN) models. Fkeras [19] is a tool helping design fault-tolerant NN models providing metrics that give a bit-level ranking of weights with respect to their sensitivity to faults. These metrics guide efficient fault injection campaigns to help evaluate the robustness of a neural network architecture and design models that are more resilient to faults. We propose to collaborate in continuing this investigation by applying Fkeras to models of interest and extending its functionalities to consider the underlying hardware implementation.

2.4 Side-Channel Mitigations

As the SoC complexity increases, the system interconnect becomes a larger target for side-channel attacks [20,21]. Many modern SoCs are not conceived to mitigate side-channel attacks. We are interested in investigating and implementing an SoC platform resilient to hardware side-channel attacks in crossbar-based and NoC architectures. Possible mitigations include traffic obfuscation to disrupt observable patterns and temporal/spatial isolation, such as partitioning and time-division arbitration. SurfNoC [10] provides a scheduler scheduling communications into "waves," mitigating side-channel attacks through non-interference among different security domains in NoC architectures. We will build upon SurfNoC to handle notions of bandwidth management and priorities that are necessary for zero-trust SoCs.

3 Key Findings, Discussion, and Significance

We aim to build a zero-trust SoC platform raising the security and safety bar in edge critical applications. Our solution is based on: (i) a zero-trust system interconnect, (ii) hardware-level real-time analysis and predictability enhancements,

(iii) hardware fault tolerance, and (iv) mitigations for side-channel attacks. Such a zero-trust platform would enable, for instance, modern accelerators for GenAI/LLM fast data access while ensuring data integrity, confidentiality, and availability of the weights, inputs, and outputs, and strong real-time guarantees for hard real-time tasks. Several of the technologies proposed in this proposal have been recently showcased for effectiveness in realistic ML mixed-critical scenarios deployed on FPGA SoCs [6,9,12] and enforce security and safety in the Carfield SoC [7]. Our proposed solutions are configurable – the impact on area/performance can be tailored to satisfy a target application's performance and area requirements. In our tested scenarios on FPGA [6,12] and ASIC [8], we experimentally measured the throughput impact to always be less than 1% and the overall increase in area less than 3%. We expect similar numbers when integrating into other platforms. The proposed functionalities extend the monitoring and control of the system, allowing the HWRoT to promptly: (i) monitor at runtime the resources executing in the system; (ii) adapt at runtime the configuration of the system according to a target execution mode, current requirements of the workload, etc.; (iii) promptly detect and react to security attacks and unsafe conditions capable of endangering the safe execution of the system. Some of the enabled features are:

- Fine-grained runtime management of the bandwidth assigned to each controller in the access of each peripheral;
- Enforcement of nominal predictable bus access to the shared interconnect;
- Prompt detection of malicious/misbehaving controllers;
- Prompt detection of unsafe conditions at the peripherals;
- Proactive mitigation of DoS attacks/conditions generated by malicious and misbehaving controllers;
- Runtime isolation of one or multiple controllers, for instance, during identified critical operations (e.g., system update, changing operation mode, etc.)
- Strong timing predictability for hard real-time tasks;
- Predictable temporary data access (DCSPM scratchpad);
- Fault tolerance in on-chip communications;
- Fault tolerance analysis for neural network models;
- Mitigation against side-channel attacks;

We aim to showcase the proposed technologies on a TRL4 demonstrator deployed on FPGA and ASIC, demonstrating the proposed functionalities on a realistic critical application deployed on the AlSaqr SoC or another SoC platform of interest.

The proposed methodologies already support crossbar-based systems, such as AlSaqr. We are currently extending them to Network-on-Chip architectures and plan to extend them to Chiplet-based architectures. These will open new interesting research opportunities at the design, analysis, and verification levels. We are also interested in enhancing our technologies and analysis for security and real-time leveraging modern LLMs and generative AI. This can be particularly interesting for analyzing tradeoffs and corner cases in complex NoC and Chiplet-based architectures.

References

1. Valente, L., et al.: A heterogeneous RISC-V based SoC for secure nano-UAV navigation. In: IEEE Transactions on Circuits and Systems I: Regular Papers, pp. 1–14 (2024)
2. Valente, L., et al.: Shaheen: an open, secure, and scalable RV64 SoC for autonomous nano-UAVs. In: 2023 IEEE Hot Chips 35 Symposium (HCS). IEEE Computer Society, Los Alamitos, CA, USA, pp. 1–12 (2023). https://doi.ieeecomputersociety.org/10.1109/HCS59251.2023.10254698
3. The AlSaqr Official GitHub Repository, TII. https://github.com/AlSaqr-platform
4. Restuccia, F., Pagani, M., Biondi, A., Marinoni, M., Buttazzo, G.: Is your bus arbiter really fair? Restoring fairness in AXI interconnects for FPGA SoCs. ACM TECS **18**(5s), 51 (2019)
5. Restuccia, F., Biondi, A., Marinoni, M., Buttazzo, G.: Safely preventing unbounded delays during bus transactions in FPGA-based SoC. In: 2020 IEEE 28th Annual International Symposium on Field-Programmable Custom Computing Machines (FCCM). IEEE (2020)
6. Restuccia, F., Kastner, R.: Cut and forward: safe and secure communication for FPGA system on chips. IEEE Trans. Comput. Aided Des. Integr. Circuits Syst. **41**(11), 4052–4063 (2022)
7. The Carfield Official Github Repository, ETH Zurich and University of Bologna. https://github.com/pulp-platform/carfield
8. Benz, T., et al.: Axi-realm: a lightweight and modular interconnect extension for traffic regulation and monitoring of heterogeneous real-time SoCs. In: Design, Automation & Test in Europe Conference & Exhibition (DATE), pp. 1–6. IEEE (2024)
9. Valente, L., Restuccia, F., Rossi, D., Kastner, R., Benini, L.: Top: towards open & predictable heterogeneous SoCs. arXiv preprint arXiv:2401.15639 (2024)
10. Wassel, H.M.G., et al.: SurfNoC: a low latency and provably non-interfering approach to secure networks-on-chip. In: Proceedings of the 40th Annual International Symposium on Computer Architecture, ser. Association for Computing Machinery, ISCA '13. New York, NY, USA, pp. 583–594 (2013)
11. AMBA®AXI™ and ACE™ Protocol Specification, ARM, IHI 0022K
12. Restuccia, F., Biondi, A., Marinoni, M., Cicero, G., Buttazzo, G.: Axi hyperconnect: a predictable, hypervisor-level interconnect for hardware accelerators in FPGA SoC. In: 57th ACM/IEEE Design Automation Conference (DAC), pp. 1–6. IEEE (2020)
13. The OpenTitan official repo, OpenTitan. https://github.com/lowRISC/opentitan
14. Restuccia, F., Pagani, M., Biondi, A., Marinoni, M., Buttazzo, G.: Modeling and analysis of bus contention for hardware accelerators in FPGA SoCs. In: 32st Euromicro Conference on Real-Time Systems (ECRTS 2020) (2020)
15. Restuccia, F., Pagani, M., Biondi, A., Marinoni, M., Buttazzo, G.: Bounding memory access times in multi-accelerator architectures on FPGA SoCs. IEEE Trans. Comput. **72**(1), 154–167 (2022)
16. Bertozzi, D., Benini, L., De Micheli, G.: Low power error resilient encoding for on-chip data buses. In: Proceedings 2002 Design, Automation and Test in Europe Conference and Exhibition, pp. 102–109 (2002)
17. Bertozzi, D., Benini, L., De Micheli, G.: Error control schemes for on-chip communication links: the energy-reliability tradeoff. IEEE Trans. Comput. Aided Des. Integr. Circuits Syst. **24**(6), 818–831 (2005)

18. Murali, S., Theocharides, T., Vijaykrishnan, N., Irwin, M., Benini, L., Micheli, G.: Analysis of error recovery schemes for networks on chips. IEEE Des. Test Comput. **22**(5), 434–442 (2005)
19. Weng, O., et al.: Fkeras: a sensitivity analysis tool for edge neural networks. J. Auton. Transp. Syst. **1**(3), 1–27 (2024)
20. Dai, M., Paccagnella, R., Gomez-Garcia, M., McCalpin, J., Yan, M.: Don't mesh around: side-channel attacks and mitigations on mesh interconnects. In: 31st USENIX Security Symposium (USENIX Security 22). USENIX Association, Boston, MA, pp. 2857–2874 (2022). https://www.usenix.org/conference/usenixsecurity22/presentation/dai
21. Paccagnella, R., Luo, L., Fletcher, C.W.: Lord of the ring(s): side channel attacks on the CPU on-chip ring interconnect are practical. In: 30th USENIX Security Symposium (USENIX Security 21). USENIX Association, pp. 645–662 (2021). https://www.usenix.org/conference/usenixsecurity21/presentation/paccagnella

Open Access This chapter is licensed under the terms of the Creative Commons Attribution 4.0 International License (http://creativecommons.org/licenses/by/4.0/), which permits use, sharing, adaptation, distribution and reproduction in any medium or format, as long as you give appropriate credit to the original author(s) and the source, provide a link to the Creative Commons license and indicate if changes were made.

The images or other third party material in this chapter are included in the chapter's Creative Commons license, unless indicated otherwise in a credit line to the material. If material is not included in the chapter's Creative Commons license and your intended use is not permitted by statutory regulation or exceeds the permitted use, you will need to obtain permission directly from the copyright holder.

Burning Fetch Execution: A Framework for Zero-Trust Multi-party Confidential Computing

Shahin Roozkhosh[1]([✉]), Bassel El Mabsout[1], Cristiano Rodrigues[3], Patrick Carpanedo[1], Denis Hoornaert[2], Su Min Tan[1], Benjamin Lubin[1], Marco Caccamo[2], Sandro Pinto[3], and Renato Mancuso[1]

[1] Boston University, Boston, USA
{shahin,bmabsout,pfcarp21,tansumin,blubin,rmancuso}@bu.edu
[2] Technical University of Munich, Munich, Germany
{denis.hoornaert,mcaccamo}@tum.de
[3] Zero-Day Labs and UMinho, Braga, Portugal
cristiano.rodrigues@algoritmi.uminho.pt, sandro.pinto@dei.uminho.pt

Abstract. How can one tamper with data that does not exist? Motivated by this question, we present the Burning Fetch eXecution (BFX) paradigm. Data *in-use* is vulnerable, and the current focus on encrypting and/or isolating *in-use* data has fallen short. Frequently reported breaches of "secure" hardware and indispensable overhead with encryption schemes confirm that trust is the modern bottleneck. This work tackles the gap in existing safeguarding technology by avoiding byte-level decryption until it is immediately fetched by the processor, only to burn it right after. We perform on-the-fetch data decryption, immediately followed by *burning*, i.e., erasing right after processing cycles. Thus, BFX minimizes the existence of sensitive data *in-use*. BFX does not demand new processing hardware units nor requires restructuring application software. Three pillars set the BFX paradigm apart: (1) *zero-trust* multi-party confidentiality with (2) security rooted in transparency, and (3) high performance.

By tackling the root of the issue, BFX enables a zero-trust multi-partied *sharing without showing* scenarios that were previously unthinkable. We showcase the impact of the BFX in a scenario with a highly privileged cloud insider attacker present. We exercise a sensitive mission whereby a third-party cloud processes fourth-party confidential real-time data streamed by a drone swarm. To further highlight the *zero-trust* nature of BFX, we assume the inference model (code) stream-processing on swarm data to be top-secret and owned by yet another party. The unknown threat, however, is the compromised processing system (cloud) where sensitive code and data are about to be deployed by all other parties-thanks to misplaced trust.

1 The Why: A *Burning* Need

The computing landscape has evolved from a single-party processing model to a multi-party environment. Market demands for higher performance and lower costs have led to adopting economies of scale in computing, resulting in the distributed processing paradigm. In this model, computationally intensive tasks are outsourced to third-party providers like Amazon AWS or Microsoft Azure. The rise of Artificial Intelligence (AI) has further accelerated this trend. However, security technologies have not kept pace with these changes.

1.1 The Unmet Need: Securely Confidential Processing

The 2019 Capital One breach[1] exposed the personal data of over 100 million customers, an act committed by a former cloud employee who leaked data from Capital One's server hosted on AWS. This incident highlights the ongoing challenges in securing data processed by third-party facilities [1]. Generally, data exists in three forms: *at-rest* (stored), *in-transit*, and *in-use* (actively processed). Cryptographic methods offer strong protection for data *at-rest* and *in-transit*; however, they fall short of protecting data *in-use*. This hinders multi-party collaboration where sensitive data is involved. Not data *in-use* can be secured, nor can parties be trusted.

While hardware-assisted security approaches have emerged as a practical solution for Confidential Computing, numerous vulnerabilities have been unveiled affecting Intel SGX [2–7], Arm TrustZone [8–10], and AMD SEV [4,11,12], exposing the weaknesses of current leading hardware (HW) technologies. Security is arguably an overloaded multifaceted term, and the inherent complexity of HW security is not well understood even by computing system professionals [13]. With only increasing threat actors' activity for financial/political gain, the message is loud and clear: existing methods remain insufficient [14]. Unsurprisingly, entities with sensitive data are hesitant about data sharing or cloud migration. Once deployed for cloud (or edge) processing, data is essentially in strangers' hands. Even with *ideal* external protection, the "trust" issue is only exacerbated as insiders account for 60% of all cyber-attacks and 43% of data breaches [15,16] in 2023 alone.

Delays, threats, and breaches have significant costs beyond mere dollars and cents. They affect organizational integrity, collective efficiency, and, worst of all, fatal catastrophes. The missed collaboration and data-sharing opportunities remain a pressing challenge to overcome.

1.2 The Shortcomings of Current Solutions

Two dominant classes of 1) Cryptographic and 2) HW-assisted methods govern Confidential computing technology. The former includes Homomorphic Encryption and secure Multi-Party Computation (MPC), imposing a considerable computational overhead. On the other hand, HW solutions, such as enclaves, offer

---
[1] Capital One breach statement.

a more efficient and practical approach, e.g., Arm's TrustZone and Intel's SGX. Combined crypto-HW methods also come with challenges as secure enclaves are primarily available for CPUs, and support for such specialized HW remains limited[2]. This restriction poses challenges, particularly in feeding data to train neural networks (NNs) [17]. Recent studies [18] show that, for instance, on Intel's SGX, this can result in an approximately 400-fold slowdown when running image classifiers.

Implementing secure hardware enclaves demands substantial application (code/data) modifications, requiring organizations with sensitive workloads to redesign their systems or hire cryptography and kernel development experts. Additionally, exclusive enclave solutions from providers like AWS and Google Cloud lead to vendor lock-ins, posing restrictions. Existing solutions are either complexly inefficient or lack robust security for broad adoption.

1.3 The Path Forward

To enable secure processing and provide the ultimate security for all data forms with added emphasis on data *in-use* hosted in untrusted environments, we introduce Burning Fetch eXecution (BFX). BFX provides a framework where sensitive data only materializes just upon the processor's fetch and vanishes immediately after processing without a trace. BFX has the potential to redefine how industries (e.g., healthcare providers, financial institutions, and governmental entities) engage with data. It facilitates seamless *sharing without showing* among parties via automatic compliance with data management strict regulations such as the EU's GDPR, the CCPA [19], and alike [20–23]. BFX provides a flexible, software-agnostic framework that can be deployed from the edge-e.g., for securing data captured by drones-to the cloud, where it can safeguard security-critical computations involving sensitive data.

2 The What: A *Burning* Strategy

What Sets Burning Fetch eXecution Apart? Three pillars set BFX paradigm apart: (1) literal *zero-rustiness*, (2) security in transparency, and (3) high performance, all without demanding new processing hardware and thus software re-usability. To effectively grasp them, it is crucial to understand why the need for high performance *undeniably* contradicts the *zero-trust* philosophy by imposing a degree of trust on multi-party confidentiality. The answer lies around transparency, the *root of the issue*.

2.1 Understanding the Root of the Issue

As detailed in Sect. 1, the primary challenge arises from the divide among parties. By formalizing the multi-party model next, we highlight the widened divide by modern AI/ML applications' reliance on HW accelerators.

---
[2] Handful of academic research has explored their potential for usage with accelerators such as GPUs and other PCI devices.

2.2 A Truly Multi-party Model

At a high level, we classify "parties" in the following multi-party/distributed secure processing model: 1) **Data Owners** are typically concerned with sharing but keen to collect data; 2) **Data Users** are solely interested in specific results attained from processing the data, (and not the data); 3) **Data Processors**, e.g., clouds, are needed but cannot be trusted by other parties. They are neither interested in the data nor the results. HW add-on accelerators also fall under this category. 4) **Data Subjects**, e.g., civilians, are reluctant with their data collected, and 5) **Data Regulators** are entities that govern, validate, and enforce situation-specific regulations.

The complexity of system-wide *trustworthiness* only increases with the **number of parties**. For example, the Data Processor party may adopt specific HW accelerators that appear insiders to the Processor party **but** manufactured by a (potentially breached) third-party HW provider.

The overarching problem we aim to address is the compromised security during its active processing (on sensitive data *in-use*) in the system, whether edge or cloud. Within the intricately interconnected nature of modern systems, laden with numerous complex on-chip components, plain-text data is considered vulnerable solely by existing. As a response, numerous cloud/edge providers and HW manufacturers have introduced hardened on-chip security. However, these solutions 1) heavily rely on placing trust in chip/service providers and 2) limit the integrability of new HW, such as GPUs, limiting any HW-assisted acceleration. Unfortunately, even with current solutions, a reported 43% of all breaches originate from "insiders", which highlights the **truth: trust is often misplaced** (see *a note on trust* in appendix A.1).

3 The How: Burning Fetch eXecution Paradigm

How is it possible to tackle the issue at its root? To minimize trust to the maximum extent, we need to think out of the (limiting) box of "secure walls" [24–26]. We propose a paradigm shift introducing a *burning* element and an *on-demand* data fetcher to our BFX paradigm.

Burning Fetch eXecution. We offer a novel BFX model that redefines data *in-use* security by removing the need for a trusting party, a model whose strength is rooted in its transparency. At the core of our technology lies the ability to decouple the physical address seen by the processor, e.g., CPU, from its corresponding memory location. This fundamental shift in approach allows for the disassociation between the contents in memory and what's fetched to the processing HW. Our ongoing research [27–30] has unlocked the potential to logically interpose a programmable module between the processors and main memory **without** any HW system on chip (SoC) modification. Leveraged by memory-mapped semantics, transactions are re-routed through the BFX module instead of taking the conventional data path. Once fetch-originated memory requests pass through the BFX logic, they are redirected to the internal memory controller. This entire mechanism effectively crafts a secondary route to memory. It is possible to use BFX routing for detailed inspection [31], profiling [32], take action based on the traffic's characteristics [28,33–35], or to enforce Confidentiality, Integrity, and Availability even on the presence of advanced software **and** hardware attackers (Trojans) [30]. Our recent advancements opened a new class of actions including (but not limited to) *just-on-fetch*, *fetch-n-burn*, and *fetch-n-monitor*. As such, BFX is enabled with a guaranteed *on-demand burning* scheme over the fetched data.

Just-on-Fetch. With current processing architectures, data must be decrypted in memory **before** its fed to the CPU. This conventional flow, involving data transfer between the CPU and memory, exposes the decrypted data while it is *in-use*, significantly increasing the potential for security vulnerabilities [30]. In contrast, *just-on-fetch* keeps data encrypted in memory at all times, decrypting it on-the-fly **just** when the CPU **fetches** it. This decryption happens immediately before the data reaches the CPU, with a cache line granularity, minimizing the exposure of raw data and substantially reducing the attack surface.

Fetch-n-Burn. To ensure that fetched data remains confined to the processor, our model mandates immediate destruction within the processing unit, leaving no residual traces. Specifically, once processed, the data bytes are instantly (and irreversibly) "**burned**". This approach guarantees unencrypted data only lives **transiently**, preventing any possibility of it residing elsewhere in the system. By enforcing data nullification at the hardware data path level, our model enables *zero-trust* secure computation on high-end COTS processing elements, even in adversarial environments.

Fetch-n-Monitor. The BFX's temporal monitoring capabilities ensure that decrypted data fed by *just-on-fetch* technique is continuously **monitored** and protected through live authentication, with unencrypted bytes available only to a specific processor within a strictly controlled time frame. Unauthorized access and anomalies [30,31,36] can be detected in real-time, even when the entire software stack is compromised. This method enables *zero-trust* and fail-safe operations.

4 Demonstration

The real-world scenario adopted in our demo is a complex inference mission involving sensitive data being processed on a cloud device, a confidential model, and civilians (see Fig. 2 in Appendix). Involved parties are unaware of an insider attacker in the target (cloud) processor. We demo how the BFX paradigm transparently 1) prevents sensitive data leakage while 2) being able to run high-performance resource-intensive tasks. In the demo, we run a real-time inference model on a live camera feed, all within a compromised cloud device!

Figure 1 illustrates how the demo follows the formalized model in Sect. 2.2. **Data Owners** represent A) confidential images about the **Data Subject(s)** to be detected (target) and B) a proprietary (and expensive) NN model owner. Both parties must maintain exclusive ownership while collaborating on sharing their data and code. **Data Processors** emulate a (cloud) system capable of processing the real-time stream from the drone swarm (with data *in-transit* protection). The swarm transmits a video stream of the **Data Subjects**, i.e., people present in the flight zone. The drones must comply with data protection protocols, stating "the captured video should not be stored, redirected, or used for any other purposes". This task is extremely challenging in untrusted multi-party scenarios. Finally, the **Data User** is the party interested in receiving live results (yes/no) on the target's presence in the surveillance footage. This party would have coordinated the mission by hiring all previous parties. Hence, they would have agreed to comply with all mission-specific and data-privacy-related protocols passed by the **Data Regulators**. Regulations encompass the combined interests and concerns of all the parties above. We show that even with compromised parties, previously unthinkable models can be governed and validated by enforcing on-demand regulations. For demonstration purposes, we also introduce an extra party, an **Attacker**, i.e., a privileged insider with access to the Data Processor (cloud), who attempts to maliciously extract the confidential code/data *in-use*.

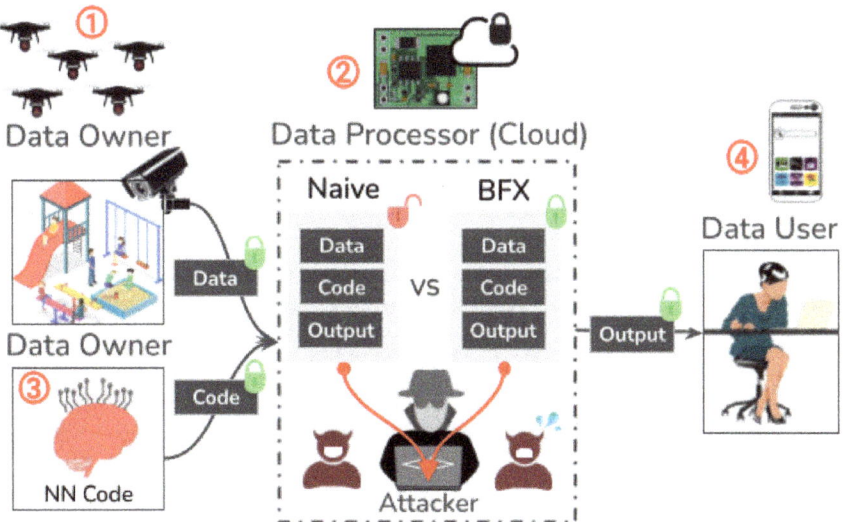

Fig. 1. The data flow between untrusted parties in the theory and demo.

Demo Setup. We consider a breached cloud provider (Data Processor) with a full-stack implementation of BFX. As depicted in Fig. 1, the cloud device ② is a scaled-down server on a Kria board. The Kria is breached by an Attacker ③, a *sudoer* who knows the exact physical addresses of sensitive code/data. The drone (**Data Owner**) captures sensitive images of people in the room (**Data Subjects**) ①. This will be depicted as a laptop with a camera transmitting an encrypted stream (i.e., data *in-transit*) to the Kria ②. For simplicity, we assume that the same Data Owner ① also securely transmits a proprietary trained NN to the Kria. The NN is written with `Tensorflow-Lite` and trained for real-time inference on the camera stream; this serves as a computationally intensive task. Finally, a mobile device (**Data User**) ④, which is solely interested in knowing if the target(s) appear(s) in the room, receiving real-time results.

We highlight the need for a *zero-trust* paradigm by allowing the attacker ③ to attempt displaying the sensitive images (demoed using a separate laptop). With our BFX implementation, the hacker only sees randomized encrypted bytes, while in the naive case, they can display the full camera stream. BFX uses authentication and phone-home procedures to perform various monitoring tasks, ensuring mission-wide system integrity.

In Conclusion. BFX achieves transparent security not by trust but by design. Adhering to the *zero-trust* philosophy, we welcome adopters to understand our design and not to trust it. With extendable support for compliant HW accelerators and adaptability to high-end COTS SoCs, BFX goes well beyond the proposed demo.

A Appendix

A.1 A Note on Trust

Schneuier's principle states that anyone can design a system they cannot break. This emphasizes the importance of independent review and transparency in system security. We aim to minimize the amount of **trust** by welcoming scrutiny. However, some form of "trust" is always involved, whether in the technology or teams. Even trusted teams may face an insider/outsider attack while not fully at fault. Lacking any trusted breach (or leak) detection technology, leakages are almost always discovered (up to months) later when the damage is permanent. The multi-faceted challenge around trust hinders actions we are already capable of performing, many of which rely on trusted data-sharing windows. We believe that trust is becoming the modern bottleneck.

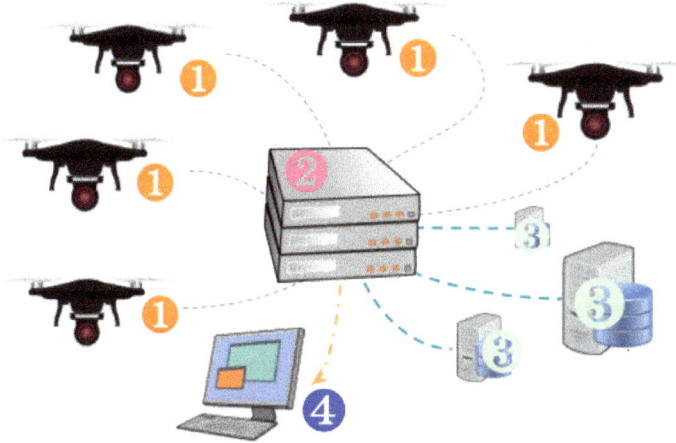

Fig. 2. A drone swarm, (1), sends data to a **Data Processor** server (2), which, in tandem, utilizes databases managed by **Data Owners** (3) to output the result to the **Data User** (4).

References

1. Khan, S., Kabanov, I., Hua, Y., Madnick, S.: A systematic analysis of the capital one data breach: Critical lessons learned. ACM Trans. Priv. Secur. **26**(1), 1–29 (2022). https://doi.org/10.1145/3546068
2. van Schaik, S., et al.: Sok: Sgx.fail: how stuff gets exposed. In: Proceedings of S&P (2024)
3. Nilsson, A., Bideh, P.N., Brorsson, J.: A survey of published attacks on intel SGX. CoRR arxiv:2006.13598 (2020)
4. Canella, C., et al.: A systematic evaluation of transient execution attacks and defenses. In: Proceedings of USENIX Security (2019)
5. Moghimi, D., Van Bulck, J., Heninger, N., Piessens, F., Sunar, B.: Copycat: controlled instruction-level attacks on enclaves. In: Proceedings of USENIX Security (2020)
6. van Schaik, S., Minkin, M., Kwong, A., Genkin, D., Yarom, Y.: Cacheout: leaking data on intel cpus via cache evictions. In: Proceedings of S&P (2021)
7. Van Bulck, J., et al.: Lvi: hijacking transient execution through microarchitectural load value injection. In: Proceedings of S&P (2020)
8. Cerdeira, D., Santos, N., Fonseca, P., Pinto, S.: Sok: understanding the prevailing security vulnerabilities in trustzone-assisted tee systems. In: 2020 IEEE Symposium on Security and Privacy (SP) (2020)
9. Rodrigues, C., Oliveira, D., Pinto, S.: Busted!!! microarchitectural side-channel attacks on the mcu bus interconnect. In: Proceedings of S&P (2024)
10. Ryan, K.: Hardware-backed heist: extracting ecdsa keys from qualcomm's trustzone. In: Proceedings of ACM CCS (2019)
11. Li, M., Zhang, Y., Wang, H., Li, K., Cheng, Y.: CIPHERLEAKS: breaking constant-time cryptography on AMD SEV via the ciphertext side channel. In: Proceedings of USENIX Security (2021)

12. Zhang, R., et al.: CacheWarp: software-based fault injection using selective state reset. In: Proceedings of USENIX Security (2024)
13. Potlapally, N.: Hardware security in practice: challenges and opportunities. In: 2011 IEEE International Symposium on Hardware-Oriented Security and Trust (2011). https://doi.org/10.1109/HST.2011.5955003
14. Gartner. Cloud adoption failures and successes (2023). https://www.gartner.com/en/doc/3987581/cloud-adoption-failures-and-successes-2023
15. Verizon. Verizon data breach investigations report. https://enterprise.verizon.com/resources/reports/dbir/
16. Securonix. 2024 insider threat report (2024). https://www.securonix.com/wp-content/uploads/2024/01/2024-Insider-Threat-Report-Securonix-final.pdf
17. Opaque. Confidential data for secure ai (2023). https://www.opaque.co/resources/articles/opaque-unveils-the-first-and-only-platform-for-running-ai-workloads-on-encrypted-data
18. El-Hindi, M., Ziegler, T., Heinrich, M., Lutsch, A., Zhao, Z., Binnig, C.: Benchmarking the second generation of intel sgx hardware. In: DaMoN '22. Association for Computing Machinery, New York (2022). https://doi.org/10.1145/3533737.3535098
19. S. of California. California consumer privacy act (2018). https://leginfo.legislature.ca.gov/faces/billTextClient.xhtml?bill_id=201720180AB375
20. Government, B.: Brazilian general data protection law. https://iapp.org/media/pdf/resource_center/Brazilian_General_Data_Protection_Law.pdf
21. Government, T.: Thailand personal data protection act (2019). https://thainetizen.org/wp-content/uploads/2019/11/thailand-personal-data-protection-act-2019-en.pdf
22. G. of India. India personal data protection bill (2019). https://prsindia.org/billtrack/the-personal-data-protection-bill-2019
23. Government, C.: Personal information protection law of the People's Republic of China. https://www.dataguidance.com/legal-research/personal-information-protection-law-peoples
24. Pinto, S., Santos, N.: Demystifying arm TrustZone: a comprehensive survey. J. ACM Comput. Surv. (2019)
25. Costan, V., Devadas, S.: Intel sgx explained (2016)
26. AMD. Protecting VM Register State With SEV-ES. AMD, Technical Report (2017)
27. Roozkhosh, S., Mancuso, R.: The potential of programmable logic in the middle: cache bleaching. In: 26th IEEE Real-Time and Embedded Technology and Applications Symposium (RTAS 2020), Sydney, Australia, April 2020, conference (2020)
28. Hoornaert, D., Roozkhosh, S., Mancuso, R.: A memory scheduling infrastructure for multi-core systems with re-programmable logic. In: Brandenburg, B.B. (ed.) 33rd Euromicro Conference on Real-Time Systems (ECRTS 2021), ser. Leibniz International Proceedings in Informatics (LIPIcs), vol. 196, pp. 2:1–2:22. Schloss Dagstuhl – Leibniz-Zentrum für Informatik, Dagstuhl (2021). https://drops.dagstuhl.de/opus/volltexte/2021/13933
29. Roozkhosh, S., et al.: Relational memory: native in-memory accesses on rows and columns. CoRR arxiv:2109.14349 (2022)
30. Rodrigues, C., Roozkhosh, S., Pinto, S., Mancuso, R.: Coherent trojans: achilles' heel of heterogeneous computing. In: Submitted IEEE S&P (2025)
31. Hoornaert, D., Roozkhosh, S., Mancuso, R., Caccamo, M.: Work in progress: identifying unexpected inter-core interference induced by shared cache. In: 2021 IEEE

27th Real-Time and Embedded Technology and Applications Symposium (RTAS), pp. 517–520 (2021)
32. Roozkhosh, S., Hoornaert, D., Mancuso, R.: CAESAR: coherence-aided elective and seamless alternative routing via on-chip FPGA. In: 43rd IEEE Real-Time Systems Symposium (RTSS) (2022)
33. Roozkhosh, S., et al.: Relational memory: native in-memory accesses on rows and columns. CoRR arxiv:2109.14349 (2021)
34. Papon, T.I., et al.: Effortless locality on data systems using relational fabric. In: Special Issue for ICDE 2023 (Best and Innovation Papers) (2023)
35. Papon, T.I., et al.: Relational fabric: transparent data transformation. In: ICDE (2023)
36. Roozkhosh, S., Hoornaert, D., Mancuso, R., Athanassoulis, M.: Hardware data re-organization engine for real-time systems. In: WiP Session @ 43rd IEEE Real-Time Systems Symposium (RTSS@Work 2022) (2022). https://par.nsf.gov/biblio/10482041

Open Access This chapter is licensed under the terms of the Creative Commons Attribution 4.0 International License (http://creativecommons.org/licenses/by/4.0/), which permits use, sharing, adaptation, distribution and reproduction in any medium or format, as long as you give appropriate credit to the original author(s) and the source, provide a link to the Creative Commons license and indicate if changes were made.

The images or other third party material in this chapter are included in the chapter's Creative Commons license, unless indicated otherwise in a credit line to the material. If material is not included in the chapter's Creative Commons license and your intended use is not permitted by statutory regulation or exceeds the permitted use, you will need to obtain permission directly from the copyright holder.

Zero-Trust Secure System and Communication Architecture to Support LLMs on the Edge Cloud Continuum (LLM-EC2)

Kanad Basu[✉] and Ifana Mahbub

Department of Electrical and Computer Engineering, University of Texas at Dallas, Dallas, TX 75252, USA
{kanad.basu,ifana.mahbub}@utdallas.edu

Abstract. Efficient adaptation of large language models (LLMs) on edge devices is crucial for applications that need ongoing, privacy-preserving adaptation and inference. This is particularly crucial for mission-critical applications, such as drones, where hardware resources are limited, and the system's overall security must operate within a zero-trust environment. In order to address these challenges, we propose a novel low-overhead hardware architecture that enables secure LLM operations on the edge. Furthermore, the proposed scheme incorporates a secure communication framework to maintain the integrity of edge operations against an external zero-trust network environment. If successful, this project will lay the foundation for a low-overhead edge environment to execute LLMs while maintaining the privacy and integrity of the system.

Keywords: Large Language Model · Hardware Accelerator · Hardware Security · Secure Communication

1 Introduction

Recently, large language models (LLMs) such as GPT-4 and LLama have demonstrated exceptional performance across various applications, significantly transforming human life. Consequently, there is a growing demand for developing efficient techniques to enable LLM operations in edge environments. However, the substantial size of LLMs poses significant challenges for adaptation on edge devices like drones. These challenges are threefold: (1) the excessive computational overhead encountered during the forward and backward passes of LLMs, (2) the substantial memory overhead required to store the massive model weights and activations, and (3) securing the LLM operations on the edge against both hardware and communication attacks when used in mission-critical systems.

Existing research on executing LLMs on the edge focuses on algorithmic-level improvements without modifying the underlying hardware. Furthermore, existing research does not take into account the privacy and security concerns of executing these LLMs on edge devices like drones. This project addresses these challenges by:

1. Designing a dedicated hardware accelerator, which reduces overall hardware overhead using parallel multiply-accumulate operations and a dedicated *softmax* module.
2. Proposing effective methods to secure the edge LLM computation against side-channel and fault attacks.
3. Securing communication using multiple sensors to control the drone's flight and the antenna aperture/beamsteering, enabling anti-jamming and interference mitigation.

The proposed platform, when ready, will be streamlined to be integrated with Al-Saqr SoC, if necessary. Both PIs are uniquely qualified to perform the proposed research. PI Basu has over 17 years of academic and industry experience in hardware design and security and has a proven track record, including joint publications with TII colleagues. PI Mahbub has been working on distributed and reconfigurable antenna arrays and secured and resilient wireless communication systems for UAVs (unmanned aerial vehicles). Her expertise will be quintessential in overcoming the challenges with swarm-based secure wireless communication platform development.

2 Proposed Approach

2.1 Task 1: Designing and Securing an LLM Hardware Engine

In this task, we propose a transformer accelerator architecture optimized for edge devices designed to support the efficient local deployment of language models while minimizing hardware resource and power consumption. Given the computational and memory demands of transformer models, specialized hardware accelerators have become essential for improving efficiency. Traditional solutions, such as GPUs and TPUs, enhance performance through parallelism and matrix operations; however, custom accelerators employ techniques like quantization and pruning to reduce complexity and resource consumption. Algorithm-hardware co-design further improves throughput and energy efficiency by aligning model optimizations with hardware constraints. A significant bottleneck in deploying large language models (LLMs) lies in memory handling, particularly in data movement between the main memory and cache, which slows down the overall process. To address this, our proposed architecture incorporates a data streaming protocol that accelerates data transfer from memory to cache while maximizing data reuse for efficient deployment on resource-constrained platforms. [4,5].

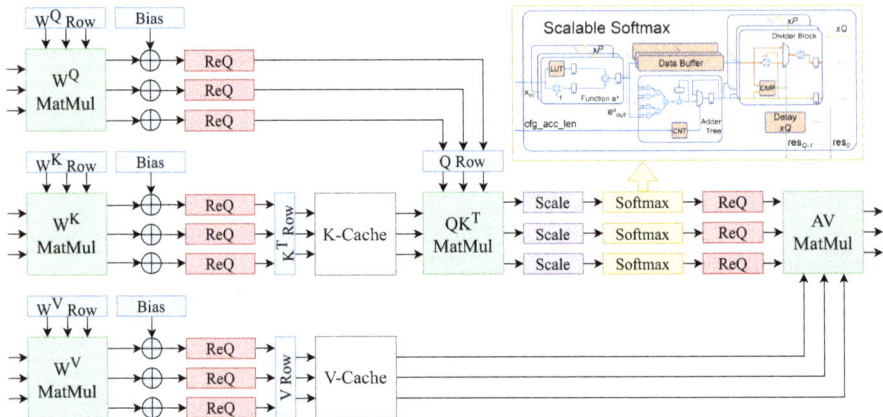

Fig. 1. Proposed transformer architecture.

Task 1.1: Proposed Transformer Accelerator Architecture. To enhance the performance of GenAI/LLMs, we aim to utilize hardware acceleration during computation and data fetch from memory. To this end, we propose a transformer accelerator that supports parallel multiplication-accumulation (MAC) operations and features an on-chip scalable *softmax* module, eliminating the need to transfer data to external compute units. Furthermore, the proposed pipelined data-feeding architecture reduces component idle time, enhancing overall efficiency. The integral components of the proposed architecture are described as follows:

Systolic Array Architecture. We propose to develop a systolic array to expedite the matrix multiplication operations, crucial for the attention mechanisms in LLMs, as depicted by *MatMul* blocks in Fig. 1. These use a grid of processor elements (PEs) that perform multiplications and additions simultaneously, holding data until the computation is complete to reduce data movement, energy consumption, and latency.

KV-Cache for Accelerated Inference. During inference, an LLM considers both the sequence of characters generated before and the current character of the prompt. Owing to the auto-regressive characteristic of the transformer-based LLM, the preceding characters have been previously computed at a specific time step. By caching the key and value matrices in a dedicated memory unit, as shown by the *K-Cache* and *V-Cache* blocks in Fig. 1, we can reduce latency by diminishing the need for recurrent computations while concurrently mitigating the computational overhead that frequent memory accesses entails [6].

Dedicated Softmax Module. We aim to utilize a dedicated softmax module for token prediction in LLMs. This module is crucial, as depicted in the *Softmax* blocks of Fig. 1. The module comprises three key components: (1) an area-efficient exponential function, (2) a partial sum accumulator, and (3) a scalable divider, as shown in the *Scalable Softmax* block. The exponential function combines a lookup table and a first-order Taylor expansion, thereby minimizing computational resource usage with only one multiplier and one adder. The partial sum accumulator, adaptable to various input vector lengths, increases the module's flexibility for different model sizes and outputs [1]. The parallel divider, consisting of cascaded subtractors and shifters, speeds up division and allows for adjustable precision. The fully pipelined softmax architecture enhances throughput and reduces latency by using hardware mechanisms tailored for the softmax function.

Integration into the SoC. The proposed transformer accelerator architecture is a novel approach designed for smaller edge devices. It focuses on compact models, like the 8-billion parameter LLaMA 3.2 model at *int8* quantization. Since this accelerator will be integrated into a SoC, the primary goal is to enhance inference performance.

One of the pivotal challenges in LLM inference is memory capacity and data transfer speed. The memory size dictates the model size that can be loaded onto the SoC; for instance, an 8B parameter model quantized to *int8* format requires approximately 8GB of memory, as each parameter occupies 1 byte. Therefore, the core focus of this study is to implement highly efficient data streaming protocols within the architecture, a crucial step toward ensuring optimal performance.

To achieve faster inference, we propose a $2048\,\text{kB} \times 2048\,\text{kB}$ systolic array inspired by Intel's NPU architecture [2]. Given the SoC bus width limitations, a dedicated cache is necessary to store data chunks that can be fed into the accelerator. A seamless data streaming method is required to fetch data from HBM or DRAM into the cache, ensuring no latency during accelerator computations. Four distinct data streaming paths need to be established: (1) Query datapath, (2) Weight datapath, (3) Key datapath, and (4) Value datapath.

The architecture must also address efficient layer-wise computation, ensuring that as one set of data completes processing, the next set is already loaded into its respective cache. This seamless, uninterrupted operation ensures a reliable and consistent performance across model layers.

Proposed Architecture and Supporting Libraries. We propose to use PCIe lanes for high-speed data transfer. Experiments on FPGA boards are necessary to determine the optimal cache size. Initially, we propose a cache size capable of storing ten rounds of the input array, which corresponds to 10 arrays of 2048 − *int*8 values, or 20kB per cache.

Running LLMs on this architecture requires support for advanced libraries, primarily PyTorch and Transformers. Initially, we will develop kernel-level support to enable accelerator usage, beginning with the RISC-V instruction set. Following this, we will implement functions that leverage the transformer accelerator architecture for necessary processes. Finally, these functions will be integrated into the PyTorch and Transformers libraries as extensions.

Evaluation Metric. To assess the accelerator's performance relative to CPUs, GPUs, and NPUs, we will examine (1) data throughput, measured by the number of tensors processed per second; (2) memory management, focusing on cache requirements compared to existing hardware and the data transfer between cache and main memory; and (3) hardware overhead.

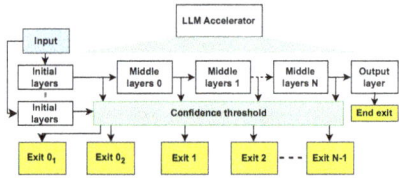

Fig. 2. Overview of the proposed fault mitigation framework.

Task 1.2: Securing the Accelerator Against Attacks. In this task, we will propose strategies to secure the accelerator designed in Task 1.1 against two hardware security vulnerabilities – side-channel and fault-attack. While side-channel attacks can compromise the confidentiality of the system, fault attacks can lead to erroneous computations, thus subverting confidence in the LLM.

Side-channel Aware Synthesis. Data communication from the SoC to the outside environment will be secured using lightweight encryption schemes. However, adversaries may launch physical side-channel attacks to subvert this encryption hardware. To this end, **we propose to perform logic synthesis for the accelerator using our side-channel aware synthesis methodology, PoSyn, which we developed with our TII colleagues** [8]. PoSyn aims to adjust the synthesis process to ensure the hardware remains secure from power side-channel (PSC) vulnerabilities, which is crucial for protecting sensitive data during LLM inference and training. It ensures resilience against PSC attacks by generating standard cell combinations tailored for PSC-vulnerable RTL components. For this purpose, it employs a bipartite matching strategy with a specialized cost function that integrates the RTL design and technology library characteristics to minimize PSC leakage. The matching process is driven by a cost function designed to reduce the distinguishability of power consumption patterns, resulting in a netlist that inherently resists PSC attacks while preserving the design's functional integrity.

Fault-attack Resilient Architecture. We propose designing efficient mitigation strategies to alleviate the impact of fault-injection attacks on large language models (LLMs) at the edge. This is achieved by leveraging shallow deep neural networks (SDNs) [3] with multiple exits, primarily used to enhance resource efficiency and reduce latency through a confidence threshold mechanism. In this work, we will extend this concept for fault detection and mitigation, which requires precise threshold estimation, unlike the original SDN design. Additionally, we intend to integrate confidence with an interlayer uncertainty thresholding mechanism to capture fault effects more accurately. The original SDN was not intended for fault detection and mitigation, making it ineffective at detecting faults in initial layers, a gap we will address in this work. These exit points, placed at various network depths, serve as redundant prediction paths to counteract fault attacks, as depicted in Fig. 2. Each exit operates independently with its own parameters and computations, ensuring isolation and preventing faults in one exit from affecting others. During inference, the system will monitor predictions from different exits and select the one unaffected by faults. Techniques such as majority voting or confidence thresholding will be employed to identify the most reliable exit.

The confidence values at multiple exits can be used to **detect** fault attacks. Normally, confidence levels improve from the initial to the final exit. Deviations from this pattern can indicate a fault. For fault-attack **mitigation**, the system can select an exit that meets a predefined confidence threshold. If no exit meets the threshold, the exit with the highest confidence will be selected. Exit isolation will ensure that faults in one exit do not affect others, and the sequential nature of exits prevents faults in middle layers from impacting earlier exits. Initial layers will be duplicated to detect faults preceding the first exit, as demonstrated in Fig. 2.

2.2 Task 2: Securing Communication

The SoC system designed in Task 1 secures the computation of LLMs, but when deployed in edge environments like drones, it is vital to safeguard communication against malicious threats, such as denial-of-service or jamming attacks. This task addresses these issues through a Generative AI (GAI)-based secure communication framework. The framework excels in managing co-channel attacks, enhancing performance in noisy environments. When jamming occurs, GAI can dynamically respond by adjusting system behavior based on interference and leveraging multiple systems for optimal communication, as illustrated in

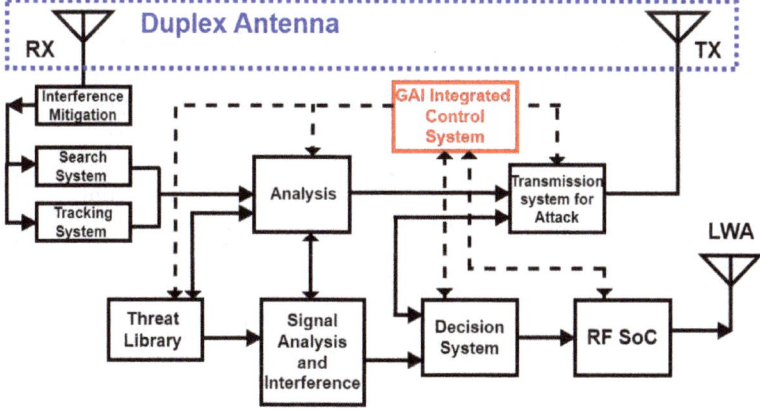

Fig. 3. Basic block diagram of an AI-based secure communication system.

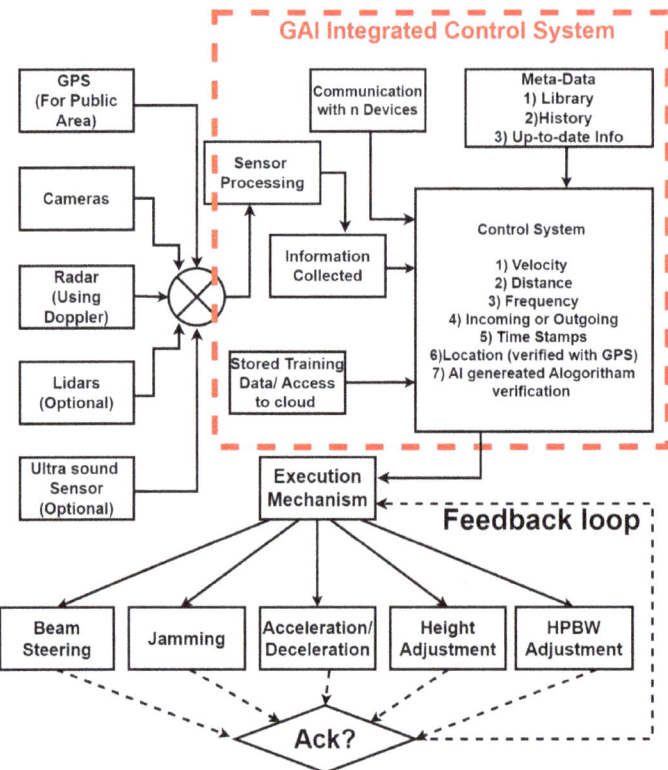

Fig. 4. Basic architecture of the drone system illustrating the data processing and communication using GAI.

Fig. 3. For LLM-based applications on edge devices, secure communication is critical. It protects the transmission of updated model parameters, preventing the exposure of sensitive information. Additionally, secure communication safeguards privacy-sensitive outputs, such as mission-critical predictions or generated reports, from interception or tampering. Integrating secure communication with the LLM hardware accelerator creates a zero-trust edge computing platform that ensures all data exchanges remain encrypted, authenticated, and verified. This unified platform is particularly relevant for drones employing LLMs for real-time decision-making or natural language processing. By establishing a zero-trust environment, the framework upholds the integrity of both LLM operations and the broader edge computing system, enabling advanced AI capabilities in challenging and potentially hostile environments. The combined security measures from Task 1 and Task 2 create a robust foundation for deploying AI in edge applications where security is paramount.

Task 2.1: Tracking and Secure Communication System. Figure 3 shows a block diagram of the proposed GAI-based communication system. Using duplex

RX and TX antennas with polarization diversity, it tracks drones by converting reflected EM waves into electrical signals for RF analysis via the Doppler effect. The system measures the signal frequency and angle of arrival (AOA) and evaluates polarization, power, pulse width, and time of arrival, storing this data in a threat library. The signal analysis block compares incoming parameters with the library to classify them as friendly or hostile. The GAI-based decision block directs the radar and RF system to continue communication with friendly drones or block it. Further, the secured communication will be carried out using a custom radio tailored for the secure communication protocol, giving full control over the firmware and ensuring seamless integration with security features. A leaky wave antenna (LWA) optimized for the TX-RX frontend will be used to enable secure frequency-hopping and beam steering for implementing a resilient communication system while enhancing processing time and spectral efficiency [7].

Figure 4 shows the proposed GAI-based flight control and communication system. The GAI-integrated controls process sensor data to manage drone operations, train drones to handle adverse weather, navigate obstacles, and maintain line of sight (LOS) communication with adjustable beamsteering. The system enables automatic flight and antenna aperture control using cloud-stored metadata and training modules for continuous threat analysis updates and performance improvements.

Task 2.2: GAI Hardware System Integration. The drone message protocol will be custom-designed for applications, including GPS, heading, timestamp, velocity, and trajectory. GAI will enhance communication with dynamic messaging and algorithmic decryption for security. **The SoC architecture from Task 1 will support GAI**, combining historical and real-time data with Zero-Trust security. The proposed system will integrate RF functions such as amplification, beamforming, and modulation. Signals are directed to an antenna array for focused communication, adding security. The control layer includes algorithms for steering, tracking, and signal analysis, with GAI improving performance and reducing latency.

Task 2.3: Interference Mitigation. In drone communication, enhancing signals from specific drones while suppressing interference is crucial. This research will use spatial filtering techniques with sensor and antenna arrays in linear, circular, or planar configurations. Key methods will include Direction of Arrival (DoA) estimation and beamforming. DoA estimation using MUSIC or ESPRIT will determine signal angles, while beamforming will adjust signal phases and amplitudes to focus only on desired signals and reject others. Both conventional (*e.g.*, delay-and-sum) and advanced (*e.g.*, MVDR) beamforming techniques will be applied. Performance will be evaluated using Signal-to-Interference-plus-Noise ratio (SINR), Bit Error Rate (BER), and signal quality, with different array configurations assessed through simulations and experiments.

3 Evaluation

We will evaluate the proposed SoC framework on six benchmarks: PubMedBert, CodeBert, T5, GPT2, LLama 3, and RoBERTa. Table 1 provides an overview of the models, detailing their respective architectures, task types, and parameter counts. The RF system evaluation measures frequency range, stability, sensitivity, output power, and signal integrity. Software and firmware will be assessed in terms of latency, accuracy, and computation power requirement. Validation and evaluation will include simulations, prototype fabrication, and field tests to confirm design accuracy and system performance. Co-PI Mahbub, in collaboration with the University of North Texas, has access to the drone flying facility, which can be utilized to perform over-the-air measurements. We will integrate the proposed framework with Al-Saqr SoC to develop a comprehensive drone solution. **The final outcome will be demonstrated on an FPGA-based platform**.

Table 1. Benchmarks used for resilience evaluation of LLMs.

Benchmark Name	Architecture	Task Type	# of Parameters
PubMedBert	Encoder-Only	Fill-mask	109.5M
CodeBert	Encoder-Only	Fill-mask	124.6M
T5	Encoder-Decoder	Translation	222.9M
Roberta	Encoder-Only	Sentiment Analysis	124.6M
GPT2	Decoder-Only	Text Completion	124.4M
LLama 3	Decoder-Only	MMLU	8B

4 Conclusion

This project proposes to develop a comprehensive low-overhead framework that secures both the computation and communication of LLMs in edge devices used in mission-critical applications, such as drones. The PIs with their interdisciplinary expertise will incorporate concepts from deep learning hardware, cybersecurity, RF communication, and security to accomplish the research objectives. If successful, this project will have a transformative impact on how deep learning systems can be implemented in edge environments, particularly drones, which can be used for various mission-critical applications, including surveillance, disaster relief, military, etc.

References

1. Fang, C., et al.: An efficient hardware accelerator for sparse transformer neural networks. In: ISCAS. IEEE (2022)
2. Intel® npu acceleration library's documentation. https://intel.github.io/intel-npu-acceleration-library/index.html. Accessed 10 Oct 2024
3. Kaya, Y., et al.: Shallow-deep networks: understanding and mitigating network overthinking. In: ICML (2019)
4. Lu, S., Wang, M., Liang, S., Lin, J., Wang, Z.: Hardware accelerator for multi-head attention and position-wise feed-forward in the transformer. In: 2020 IEEE 33rd International System-on-Chip Conference (SOCC), pp. 84–89. IEEE (2020)
5. Park, J., Yoon, H., Ahn, D., Choi, J., Kim, J.J.: Optimus: optimized matrix multiplication structure for transformer neural network accelerator. Proc. Mach. Learn. Syst. **2**, 363–378 (2020)
6. Pope, R., et al.: Efficiently scaling transformer inference. Proc. Mach. Learn. Syst. **5**, 606–624 (2023)
7. Sah, P., et al.: A 38° wide beam-steerable compact and highly efficient v-band leaky wave antenna with surface integrated waveguide for vehicle-to-vehicle communication. In: TSWMCS, pp. 1–5 (2023)
8. Srivastava, A., et al.: Posyn: a graphical approach towards side-channel aware synthesis. In: IEEE HOST (under review) (2025)

Open Access This chapter is licensed under the terms of the Creative Commons Attribution 4.0 International License (http://creativecommons.org/licenses/by/4.0/), which permits use, sharing, adaptation, distribution and reproduction in any medium or format, as long as you give appropriate credit to the original author(s) and the source, provide a link to the Creative Commons license and indicate if changes were made.

The images or other third party material in this chapter are included in the chapter's Creative Commons license, unless indicated otherwise in a credit line to the material. If material is not included in the chapter's Creative Commons license and your intended use is not permitted by statutory regulation or exceeds the permitted use, you will need to obtain permission directly from the copyright holder.

Predictive Maintenance System for Enhancing Chip Reliability and Resiliency in UxVs

Freddy Gabbay(✉)

The Institute of Applied Physics, The Hebrew University of Jerusalem, Jerusalem, Israel
freddy.gabbay@mail.huji.ac.il

Abstract. The emergence of Unmanned Vehicles (UxVs) has revolutionized various industries, offering unprecedented capabilities in areas such as surveillance, logistics, and environmental monitoring. As UxVs become increasingly integral to critical operations, the reliability of their components, particularly semiconductor chips, becomes paramount. Ensuring chip reliability is crucial for maintaining the overall performance and safety of UxVs, particularly in critical systems such as navigation, obstacle detection, environmental sensing, and data transmission. Reliability failures can be exploited to breach security, leading to unauthorized access or data corruption. In recent years, advancements in semiconductor technology have increased the vulnerability of semiconductors to reliability issues caused by deterioration over time, such as transistor aging. Aging related failures are accelerated by rising temperature, particularly emphasized in advanced technologies due to the increase thermal density. Resolving such reliability issues becomes challenging due to the complex interplay of physical mechanisms, environmental conditions, and compute workloads, each requiring adaptive mitigation strategies. This paper introduces a novel predictive maintenance system for detecting and mitigating reliability failures in UxVs. To the best of our knowledge, this is the first predictive maintenance system introduced specifically for IC reliability. The proposed system can analyze multi-dimensional sensor data to provide comprehensive and robust predictive maintenance, dynamically addressing evolving aging faults. The proposed system consists of three subsystems: A set of sensors that can identify evolving failures before they manifest in the crucial functional systems of the UxV; a set of maintenance agents that can mitigate various types of progressing failures; and a control subsystem that orchestrates the sensors and the maintenance agents. Our preliminary simulations suggests that the proposed system has the potential to mitigate such reliability failures.

Keywords: Reliability · UxVs · Predictive Maintenance · Circuit Aging

1 Introduction

In recent decades, VLSI technologies have advanced remarkably. Continuous development of new process nodes has led to the consistent miniaturization of transistors to nanometric dimensions. Additionally, revolutionary devices and materials have driven improvements in performance and energy efficiency. These technologies have enabled the development of highly sophisticated UxV systems by allowing the integration of complex functionalities into compact, efficient semiconductor chips. The miniaturization process and increased computational power have enabled enhanced capabilities in UxVs, such as real-time data processing, autonomous navigation, and advanced sensor integration, significantly improving their performance and functionality. However, these advancements have also exposed UxV ICs to reliability issues, particularly those caused by aging, resulting in over-time wear-out. Aging in integrated circuits (ICs) is primarily caused by mechanisms such as Bias Temperature Instability (BTI) [1] and Hot Carrier Injection (HCI) [2], which degrade transistor performance over time. BTI occurs when prolonged voltage stress at high temperatures leads to threshold voltage shifts, while HCI results from high-energy carriers damaging the transistor's gate oxide. For instance, about 70% of the threshold voltage degradation over a 10-year period occurs within the first year. Asymmetric transistor aging [2] can be even more complicated since it causes uneven degradation across different parts of the circuit, leading to severe timing violations and unpredictable performance issues. The accumulation of wear-out failures can propagate through the UxV's critical systems, potentially leading to overall system failure and significant safety concerns. As the reliability of key components degrades, the UxV's ability to perform essential functions such as navigation, communication, and obstacle detection is compromised, thereby increasing the risk of operational failures and accidents. Additionally, such reliability failures can be exploited by attackers to breach security, potentially leading to unauthorized access, data corruption or denial of service. These vulnerabilities can serve as entry points for malicious activities, compromising the integrity of the system.

This paper introduces a novel predictive maintenance system for UxVs that performs dynamic reliability monitoring using a specialized array of sensors and anti-aging agents which are orchestrated by a predictive maintenance control unit. To the best of current knowledge, this is one of the first proposed predictive maintenance systems specifically targeting IC reliability, representing a robust approach to addressing the significant reliability challenges of semiconductors in advanced process nodes. Employing such a predictive maintenance approach in UxVs can significantly extend their reliability, resiliency, safety and operational efficiency. By continuously monitoring the health of critical components via a specialized array of sensors, the system can robustly detect early indicators of wear-out and potential failures. The incorporation of a designated predictive maintenance control unit can provide accurate predictions for complex scenarios and enable timely interventions. Furthermore, it can process multi-dimensional data from an array of sensors and employ multiple mitigation strategies to overcome the limitations of existing approaches, which are constrained by their focus on sensing a single parameter and using a single mitigation mechanism. The anti-aging agents can be triggered to prevent unexpected reliability failures when required, thereby extending the operational lifetime of UxV ICs. Consequently, this approach improves safety, reduces maintenance costs, and increases mission success rates, underscoring its importance for the effective deployment of UxVs across diverse applications. The remainder of this paper is organized as follows: Sect. 2 describes the proposes UxVs predictive maintenance system, Sect. 3 presents our initial simulation analysis and key findings, and Sect. 4 concludes this paper.

2 Predictive Maintenance System

The proposed predictive maintenance system for UxV ICs reliability is illustrated in Fig. 1. The system consists of the following subsystems:

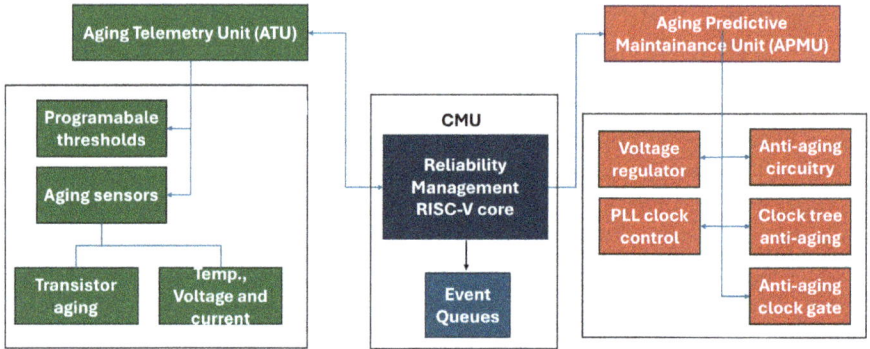

Fig. 1. UxV predictive maintenance system for chip reliability.

2.1 Control and Management Unit (CMU)

The CMU consists of a RISC-V core connected to an aging telemetry unit (ATU) which constitutes of a network of sensors distributed across the UxV ICs. The CMU dynamically tracks the wear-out of the UxV logical elements during operation, alongside predictive maintenance hardware mechanisms for addressing reliability issues detected in real-time. The core runs predictive maintenance algorithms that dynamically orchestrate the reliability of all system components. The CMU can detect reliability-related anomalies and predict wear-out failures in the field while considering system workload and environmental conditions. In addition, the CMU can perform predictive maintenance of chip health based on the data collected by the sensors. When required, it will trigger the necessary mitigation mechanisms to prevent such runtime failures. The CMU is equipped with event queues (EQs) that serve for reporting reliability-related events to the system operator or a system security management unit. The EQs can capture both informative events and critical incidents requiring external intervention. For example, the EQs can report reliability events that highlight potential vulnerabilities, as these events could reveal eligibility-related issues that might be exploited for cyber-attacks and security breaches if not properly secured.

2.2 Aging Telemetry Unit (ATU)

The ATU consists of a suite of aging sensors that can offer insights into the UxV chips' health. The sensors are equipped with programmable thresholds to define the acceptable operational range. Should any measurement fall outside these set boundaries, it will be promptly reported to the CMU. The sensors will possess the capability to identify various potential wear-out related failure causes, either before they manifest or as they occur. The data collected by the ATU will be transmitted to the predictive maintenance applications in the CMU, which will analyze the information to determine the necessary actions for mitigating runtime reliability failures. The ATU can be equipped with the following set of sensors:

Ring Oscillators (ROs): ROs are circuits composed of an odd number of logic inverters arranged in a closed loop [3], which results in a signal oscillating between logical high and low states. Figure 2 illustrates a schematic of two ROs each using 51 inverters or 51 NAND gates. The oscillation frequency of an RO is affected by the number of logic elements and the propagation delay of each component. ROs provide a compact and consistent design that can be efficiently deployed across multiple locations on a chip to monitor BTI degradation over time.

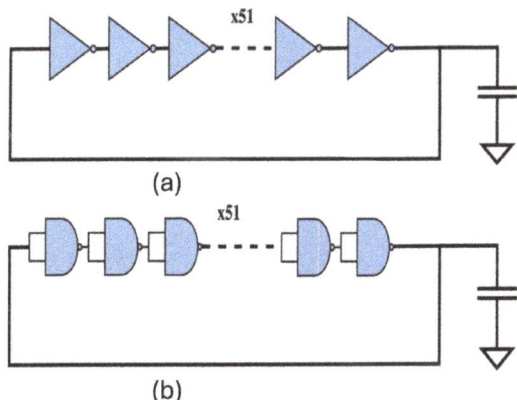

Fig. 2. Ring oscillator circuits with: (a) 51 Inverters, and (b) 51 NAND gates.

Clock delay shift telemetry sensor: The clock delay shift telemetry sensor [4] scheme is illustrated in Fig. 3. The sensor is utilized to measure the delay shift due to asymmetric transistor aging [2] in the clock tree between two chip modules. The clock network in the illustrated figure comprises launch and capture branches connected to modules A and B, respectively. Each branch is typically managed by a clock gate to optimize dynamic power consumption. When the clock is gated, BTI stress is introduced to the clock tree buffers. An imbalance in the clock gating ratio between the launch and capture branches can result in asymmetric transistor aging, which may lead to both setup and hold timing violations. The sensor comprises of the following elements:

1. Programmable Delay Circuit (PDC): The PDC consists of a series of pass-transistor multiplexers. It allows for precise adjustment of the delay from the input, DataIn,

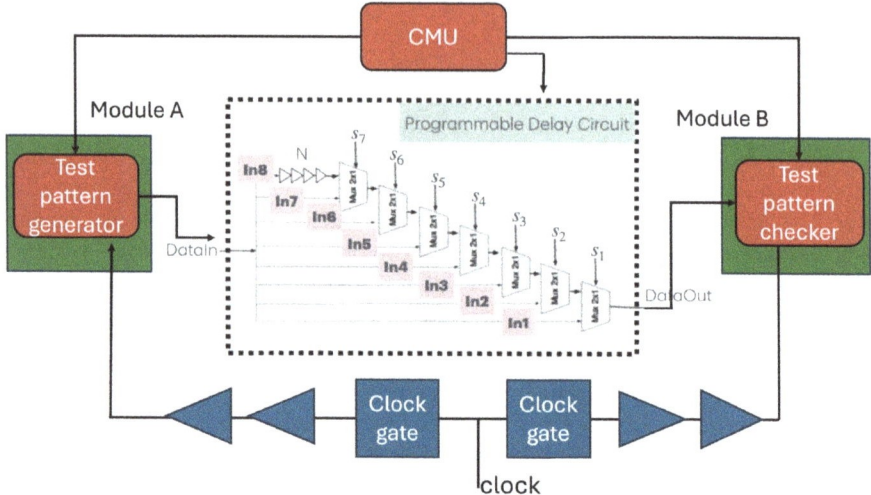

Fig. 3. Telemetry circuitry for measuring clock tree timing deterioration due to asymmetric transistor aging.

to the output, DataOut, via control signals (S1-Sm), where m is the number of multiplexers (illustrated in Fig. 3 for m = 7). The use of pass-transistor multiplexers enhances control and precision, enabling finer delay increments, which is essential for minimizing clock skew.

2. Test Pattern Generator (TPG): The TPG includes a pseudo-random bit sequence (PRBS) generator [4], which deterministically produces binary sequences that mimic randomness.
3. Test Pattern Checker (TPC): The TPC verifies the integrity of received PRBS data by comparing it to the expected sequence. Detected errors are recorded in an error counter.

The telemetry circuitry is managed by the CMU, which periodically activates the TPG and TPC while sweeping the PDC delays in an incremental manner. This procedure extracts the available maximum and minimum delay margins for setup and hold timing constraints, respectively.

Built-In Self-test: Logic Built-In Self-Test (LBIST) and Memory Built-In Self-Test (MBIST) circuitries are commonly used to enhance testability and fault detection in ICs. LBIST and MBIST can be activated to perform periodic chip health checks to detect potential deterioration, which will be reported to the CMU.

2.3 Aging Telemetry Unit (ATU)

The APMU is controlled by CMU that dynamically trigger its activation upon identifying potential failures. The APMU encompasses mechanisms for both failure prevention and correction. Prevention circuits include various modules, such as transistor anti-aging circuitry to counteract asymmetric transistor aging, and clock tree anti-aging circuitry, among others. Dynamic correction mechanisms include controls to adjust the PLL clock frequency and regulate the supply voltage to compensate for aging degradation. To illustrate the operation of the APMU, two anti-aging circuitries illustrated in Fig. 4:

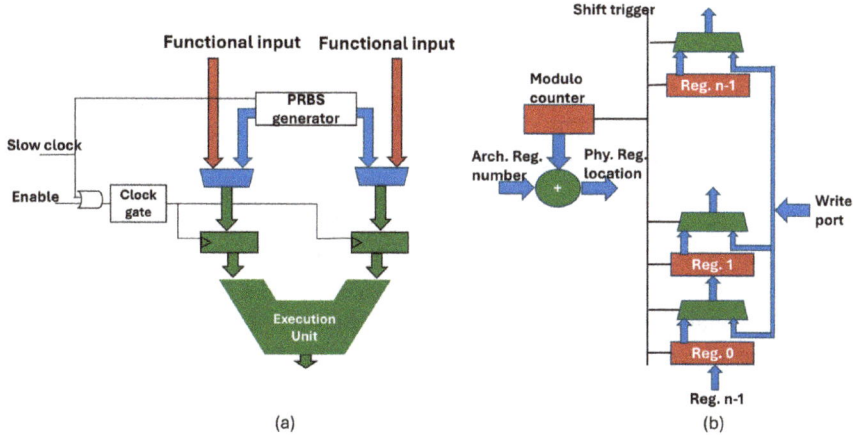

Fig. 4. Anti-aging circuitries for (a) data path modules and (b) register files.

Data Path Anti-aging Agent: The first anti-aging circuitry, designed for data path logical structures [2], is illustrated in Fig. 4(a). It uses a PRBS generator activated by a low-frequency clock to produce pseudorandom patterns, which are injected into idle execution units and data path structures to prevent extended periods of constant stress that can intensify BTI. The activation of the agent is controlled by the CMU when required. The agent employs isolation cells [5] (similar to their use in design-for-testability and power islands) to ensure that it operates seamlessly without affecting the surrounding components.

Register Files (RFs) Anti-aging Agent: The RF anti-aging circuitry [2], illustrated in Fig. 4(b), periodically changes the mapping of architectural registers to their corresponding physical hosting locations. The scheme is based on modulo rotation of the mapping between the architectural register identifiers and their physical locations. A pulse trigger is asserted to shift the register mapping in RF periodically at low frequency controlled by the CMU. A modulo-counter serves to map the architectural register number to the physical register location by modulo addition. After each assertion of the rotation trigger, the counter is incremented, and the register values are shifted between registers. The proposed circuitry can eliminate both long-duration idle states in registers that are under-utilized, thereby helping to mitigate the long-term impact of BTI.

3 Simulation Analysis and Key Findings

The simulation analysis and key findings cover both aging sensors and anti-aging agents described in Sect. 2. The circuits were designed using a 28 nm process node, assuming a slow-slow process corner for timing analysis, a core voltage of 0.9 V, and a junction temperature of 125 °C. The simulation environment uses the Cadence Spectre® circuit-level simulator [6] with OMI aging libraries [7], which model the impact of BTI on N-type and P-type transistors across various V_{th} values, assuming a 10-year lifetime.

3.1 Transistor Aging Sensors

The first simulation results, presented in Fig. 5, illustrate the BTI effect on ROs designed using three different types of threshold voltage transistors: Standard V_{th} (SVT), Low-V_{th} (LVT), and ultra-Low V_{th} (uLVT). The initial observation is that P-type transistors exhibit greater aging degradation than N-type transistors, consistent with previous studies [1]. Figure 5(a) shows that SVT transistors, both N-type and P-type, have a higher percentage of saturation current, I_{d-sat}, degradation compared to LVT and uLVT transistors. The saturation current is the maximum current that flows through the transistor when it is in the saturation region, where the current becomes constant despite increases in the drain-source voltage. Figure 5(b) shows that SVT transistors also experience the most significant frequency degradation in NAND- and inverter-based ROs, reaching about 8.7% in the worst case, whereas LVT and uLVT transistors show lower degradation of 7% and 6.2%, respectively.

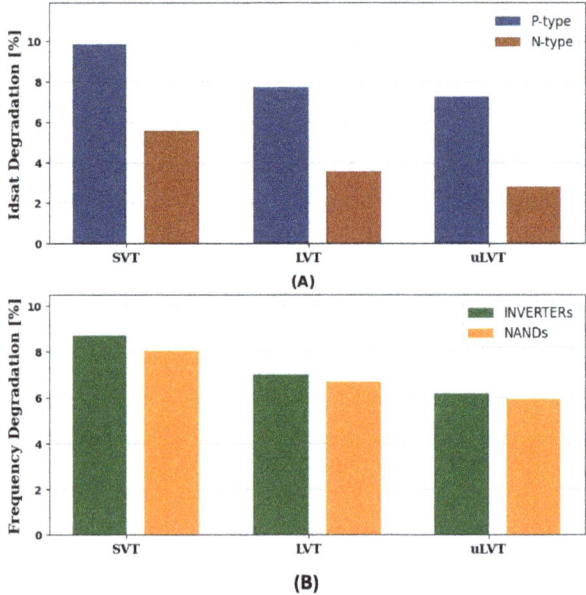

Fig. 5. Multi-Vth comparison: Percentages of (A) Id − sat degradation, (B) frequency degradation comparing inverters and NAND-based ROs.

The next simulation analysis demonstrates the clock tree timing sensor using a case study with two different implementations for the clock tree branches, assuming LVT clock inverters and a 1 ns insertion delay in the fresh state: 1) a branch with cells only, where wire delays due to resistance and capacitance (RC) are negligible, and 2) a mixed

RC-and-cell branch, where 30% of the clock branch insertion delay is due to wire delay, which is not affected by BTI. Our simulation analysis demonstrates the capability of the circuit for detection of setup timing violations assuming cell-only branch for clock launch (clk1) and the RC-and-cell mixed branch for capture clock (clk2). In this example, at t = 0, with a fresh device, the maximum allowed delay to meet setup constraints is achieved when the PDC selectors are set to $(0001111)_2$, corresponding to a delay of the chain IN5 in the PDC. When the circuitry is aged, clock skew shifts may occur. As long as no timing failures are detected with the PDC configuration used in t = 0, the PDC selector is unchanged. Once a violation is detected, the PDC selectors are swept to identify the longest possible delay chain that exhibits no violations. Figure 6 presents a shmoo chart illustrating the detection capabilities of the sensors under different BTI stress conditions. Each column and row represent a different BTI stress duration for the launch and capture clocks, respectively. Each cell in the table indicates the maximum possible delay chain through which the PRBS data can be transferred without error. Gray cell colors represent BTI stress combinations on launch and capture clocks where the telemetry circuitry can successfully pass the PRBS, while the red cells indicate combinations of stress that fail. The different shades of gray represent gradual setup slack degradation as the PDC delay is decreased among the shaded groups. Similarly, the sensor can be used to detect hold slack degradation by sweeping the PDC and identifying the minimum delay required for the circuit to operate without failures.

CLK1 / CLK2	2%	5%	10%	25%	50%	75%	90%	98%
2%	IN4	IN3	IN2	IN1	X	X	X	X
5%	IN4	IN3	IN3	IN2	IN1	X	X	X
10%	IN4	IN4	IN3	IN3	IN2	IN1	X	X
25%	IN5	IN4	IN4	IN3	IN3	IN2	IN2	IN2
50%	IN5	IN5	IN4	IN4	IN3	IN3	IN2	IN2
75%	IN5	IN5	IN5	IN4	IN4	IN3	IN3	IN3
90%	IN5	IN5	IN5	IN4	IN4	IN3	IN3	IN3
98%	IN5	IN5	IN5	IN4	IN4	IN3	IN3	IN3

Fig. 6. Shmoo chart – Setup failure detection.

3.2 Anti-aging Agents

The anti-aging circuits for the data path modules and RFs have been examined on a case-study of a microprocessor floating-point (FP) adder and an RF. Table 1 summarizes the timing analysis where it demonstrates that significant timing violations can arise due to BTI and asymmetric aging. In addition, the results indicate both anti-aging agents have successfully eliminated the violations observed.

Table 1. Aging-Aware Timing Analysis.

Functional Units	Timing Mode	Number of Violations/Total Paths		
		Fresh	Aged	Anti-Aging agents
FP Adder	Setup	0/32598	69/32598	0/32598
RF	Setup	0/215040	132/215040	0/215040
FP Adder	Hold	0/32598	3/32598	0/32598
RF	Hold	0/215040	0/215040	0/215040

4 Simulation Analysis and Key Findings

The emergence of UxVs has revolutionized various industries, offering unprecedented capabilities in various mission critical areas. As UxVs become increasingly integral to critical operations, ensuring the reliability of their onboard semiconductor chips is crucial for maintaining overall resiliency and safety. This paper introduces a novel predictive maintenance system for detecting reliability failures in UxVs. The proposed approach comprises three subsystems: a set of sensors to identify evolving failures before they impact critical UxV functions, a set of agents to mitigate various types of progressing failures, and a control subsystem that orchestrates the overall system. Preliminary simulation analysis indicates that the system has the potential to mitigate reliability failures, thereby enhancing the safety and resiliency of UxVs.

References

1. Alam, M.A., Mahapatra, S.: A comprehensive model for PMOS NBTI degradation. Microelectron. Rel. **47**(6), 853–862 (2007)
2. Gabbay, F., Mendelson, A.: Asymmetric aging effect on modern microprocessors, Microelectronics Reliability, Vol. 119 (2021)
3. Gabbay, F., Ella, M., Ramadan, F., Wattad, D.: An analysis of BTI-induced degradation on multi-Vth 28-nm Ring Oscillator. In: Proceedings of the 6th International Conference on Microelectronic Device and Technologies (MicDAT '2024)
4. Gabbay, F., Wattad, D.: On-die telemetry circuitry for measuring clock tree timing deterioration due to asymmetric transistor aging. In: Proceedings of the 6th International Conference on Microelectronic Device and Technologies (MicDAT '2024)
5. Agrawal, V.D., Cheng, K.-T., Johnson, D.D., Sheng Lin, T.: Designing circuits with partial scan. IEEE Design Test Comput. **5**(2), 8–15 (1988). https://doi.org/10.1109/54.203
6. Cadence Virtuoso Spectre Circuit Simulator User Guide
7. Davis, W.R., Shaw, C., Hassan, A.R.: How to write a compact reliability model with the Open Model Interface (OMI). In: 2020 IEEE International Reliability Physics Symposium, USA, pp. 1–2 (2020)

Open Access This chapter is licensed under the terms of the Creative Commons Attribution 4.0 International License (http://creativecommons.org/licenses/by/4.0/), which permits use, sharing, adaptation, distribution and reproduction in any medium or format, as long as you give appropriate credit to the original author(s) and the source, provide a link to the Creative Commons license and indicate if changes were made.

The images or other third party material in this chapter are included in the chapter's Creative Commons license, unless indicated otherwise in a credit line to the material. If material is not included in the chapter's Creative Commons license and your intended use is not permitted by statutory regulation or exceeds the permitted use, you will need to obtain permission directly from the copyright holder.

Author Index

A
Abernethy, Jacob 147
Acquaviva, Andrea 131
Alhammadi, Yousof 32
Alhussein, Omar 38
Ali, Abubakar S. 38
Almemari, Al Anoud 32
Anisetti, Marco 81
Ardagna, Claudio A. 81
Arif, Muhammad Shahzad 138
Asal, Rasool 32
Atrouz, Mohammad 165
Azriel, Leonid 98
Azzouni, A. 21

B
Barbosa, Diego Ortiz 12
Barbosa, Guilherme Nunes Nasseh 124
Barchi, Francesco 131
Baskin, Chaim 106
Bastoni, Andrea 131
Basu, Kanad 206
Bena, Nicola 81
Burbano, Luis 12
Byeon, Sooyung 155

C
Caccamo, Marco 196
Cao, Yinzhi 12
Cardenas, Alvaro A. 12
Carpanedo, Patrick 196
Chen, Ang 47
Choi, Joonwon 155
Colombo, Maurizio 32
Costa, Miguel 131, 181
Cully, Antoine 90

D
Damiani, Ernesto 32, 38, 81

E
El Mabsout, Bassel 196
Esposo, Chris 147

G
Gabbay, Freddy 216
Ganesh, Vijay 147
Ghadban, Adam 98
Gomes, Tiago 181
Goppert, James M. 3
Guo, Yifan 3

H
Haneefa, Fayaz Mohamed 165
Hoornaert, Denis 196
Horyna, Jiri 66
Hwang, Inseok 3, 155
Hwang, Sounghwan 3

K
Kastner, Ryan 188
Krishnamachari, Bhaskar 56

L
Lee, Wenke 147
Licea, Daniel Bonilla 117
Lubin, Benjamin 196

M
Mahbub, Ifana 206
Mancuso, Renato 196
Mattos, Diogo Menezes Ferrazani 124
Mendelson, Avi 98, 106
Michalas, Antonis 72
Muhaidat, Sami 38, 138

N
Naser, Shimaa 38
Nemcovsky, Yaniv 106

P

Pan, Leyan 147
Pant, Kartik A. 3
Parisi, Emanuele 131
Pellizzoni, Rodolfo 131
Penicka, Robert 172
Pinto, Sandro 131, 181, 196
Pujolle, G. 21

Q

Qassem, Lamees M. Al 32

R

Rabaninejad, Reyhaneh 72
Restuccia, Francesco 188
Rodrigues, Cristiano 196
Roozkhosh, Shahin 196
Rossi, Davide 188

S

Sadia, Mushtari 47
Sarkar, Tamoghna 56

Saska, Martin 66, 117, 172
Shoufan, Abdulhadi 165
Silano, Giuseppe 117
Sofotasios, Paschalis C. 138

T

Tan, Su Min 196

V

Vijay, Vishnu 3

W

Wang, Zijun 12

X

Xiao, Yunming 47
Xie, Cihang 12

Y

Yang, Siwei 12
Yeun, Chan Yeob 81
Yoon, Sangyoung 81

Open Access This book is licensed under the terms of the Creative Commons Attribution 4.0 International License (http://creativecommons.org/licenses/by/4.0/), which permits use, sharing, adaptation, distribution and reproduction in any medium or format, as long as you give appropriate credit to the original author(s) and the source, provide a link to the Creative Commons license and indicate if changes were made.

The images or other third party material in this book are included in the book's Creative Commons license, unless indicated otherwise in a credit line to the material. If material is not included in the book's Creative Commons license and your intended use is not permitted by statutory regulation or exceeds the permitted use, you will need to obtain permission directly from the copyright holder.

GPSR Compliance

The European Union's (EU) General Product Safety Regulation (GPSR) is a set of rules that requires consumer products to be safe and our obligations to ensure this.

If you have any concerns about our products, you can contact us on

ProductSafety@springernature.com

In case Publisher is established outside the EU, the EU authorized representative is:

Springer Nature Customer Service Center GmbH
Europaplatz 3
69115 Heidelberg, Germany

www.ingramcontent.com/pod-product-compliance
Lightning Source LLC
Chambersburg PA
CBHW071744271025
34597CB00003B/182